# S&T Diplomacy and Sustainable Development in the Developing Countries

## About the Centre

The Centre for Science and Technology of the Non-Aligned and Other Developing Countries (NAM S&T Centre) is an inter-governmental organisation with a membership of 48 countries spread over Asia, Africa, Middle East and Latin America. Besides this, 12 S&T agencies and academic/research institutions of Bolivia, Botswana, Brazil, India, Nigeria and Turkey are the members of the S&T-Industry Network of the Centre. The Centre was set up in 1989 to promote South-South cooperation through mutually beneficial partnerships among scientists and technologists and scientific organisations in developing countries. It implements a variety of programmes including international workshops, meetings, roundtables, training courses and collaborative projects and brings out scientific publications, including a quarterly Newsletter. It is also implementing 7 Fellowship schemes, namely, NAM S&T Centre Research Fellowship, Joint NAM S&T Centre – ICCBS Karachi Fellowship, Joint CSIR/CFTRI (Diamond Jubilee) - NAM S&T Centre Fellowship, Joint NAM S&T Centre – ZMT Bremen Fellowship, Research Training Fellowship for Developing Country Scientists (RTF-DCS), NAM S&T Centre – U2ACN2 Research Associateship in Nanosciences and Nanotechnology and Joint NAM S&T Centre – DST (South Africa) Training Fellowship on Minerals Processing and Beneficiation in Indian institutions. These activities provide, among others, the opportunity for scientist-to-scientist contact and interaction, training and expert assistance, familiarising the scientific community on the latest developments and techniques in the subject areas, and identification of technologies for transfer between member countries. The Centre has so far brought out 70 publications and has organised 100 international workshops and training programmes.

For further details, please visit www.namstct.org or write to the Director General, NAM S&T Centre, Core 6A, 2nd Floor, India Habitat Centre, Lodhi Road, New Delhi-110003, India (Phone: +91-11-24645134/24644974; Fax: +91-11-24644973; E-mail: namstcentre@gmail.com; namstct@bol.net.in).

# S&T Diplomacy and Sustainable Development in the Developing Countries

*— Editors —*

**Dr. Tahereh Miremadi**

**Mr. Abdul Haseeb Arabzai**

**Mrs. Sadhana Relia**

CENTRE FOR SCIENCE & TECHNOLOGY OF THE
NON-ALIGNED AND OTHER DEVELOPING COUNTRIES
(NAM S&T CENTRE)

2017
DAYA PUBLISHING HOUSE®
*A Division of*
ASTRAL INTERNATIONAL PVT. LTD.
New Delhi – 110 002

ISBN: 978-93-86071-50-7 (International Edition)

*Publisher's Note:*

Centre for Science and Technology of the Non-Aligned and Other Developing Countries (NAM S&T Centre)
Core-6A, 2nd Floor, India Habitat Centre, Lodhi Road,
New Delhi-110 003 (India)
Phone: +91-11-24644974, 24645134, Fax: +91-11-24644973
E-mail: namstct@gmail.com
Website: www.namstct.org

*Published by*      :   **Daya Publishing House®**
                        *A Division of*
                        **Astral International Pvt. Ltd.**
                        – ISO 9001:2015 Certified Company –
                        4736/23, Ansari Road, Darya Ganj
                        New Delhi-110 002
                        Ph. 011-43549197, 23278134
                        E-mail: info@astralint.com
                        Website: www.astralint.com

Former Chairman, Overseas
Infrastructure Alliance
(India) Pvt. Ltd.
B-20, Pamposh Enclave-I, New
Delhi-110048, India

*Ambassador V. B. Soni*
I.F.S. (RETD.)

Tel: +91-11-41074202
Mob: +91-9811699191
E-mail: vbsoni@hotmail.com
vbsoni13@gmail.com

# Foreword

Almost fifty years ago when I joined the Indian Foreign Service, we learnt that diplomacy was the instrument of safeguarding one's country's national interest to achieve its goals through peaceful cooperation and negotiation with other nations. In recent times, due to the deepening level of globalization and transnational activities, diplomacy's whole definition has changed. The emphasis is no more restricted to political, military, strategic, commercial/economic, or cultural issues. With the entry of science and technology, a totally new and exciting area has opened up of international cooperation which would be of particular interest to developing countries.

Science can advise and support foreign policy objectives. Diplomacy can facilitate international scientific cooperation. Scientific cooperation can improve international relations between countries. An expert recently observed that: "Beyond providing knowledge and applications to benefit human welfare, scientific cooperation is a useful part of diplomacy - scientific cooperation to work on problems across borders and without boundaries, cooperation made possible by the international language and methodology of science, cooperation in examining evidence that allows scientists to get beyond ideologies and form relationships that allow diplomats to defuse politically explosive situations."

I am of the firm belief that many global challenges related to health, food security, economic growth, and climate change lay at the intersection of science and international relations.

In the present knowledge-based global economy, Science and Technology (S&T) competitiveness is becoming critical for long term sustainable development of any country. With the rapid advances in science and technology, more and more challenges can be faced by nations successfully. The role of governments is becoming critical and countries are required to respond through foreign policy programmes, institutions, and diplomatic partnerships to maximise their economic opportunities and facilitate finding solutions to the problems of the partnering nations.

Science as a tool for diplomacy has been used for several decades and by many countries around the world. One of the earliest ventures in joint scientific cooperation was in 1931 with the creation of the International Council of Scientific Unions, now the International Council for Science (ICSU), to strengthen international science for the benefit of society by mobilising the knowledge and resources of the international science communitytowards the further developments and scientific solutions to the world's challenges. Science diplomacy is thus a tool for bilateral/multilateral/regional cooperation for promotion of relations through S&T cooperation, extending to mutual trade and commerce.

At present, the world is facing several challenges, such as ensuring food security, supplying clean water, battling infectious diseases, mitigating climate change, addressing urbanisation, building green energy economies, and reducing biodiversity loss, all of which need considerable transformational and innovative solutions with the participation of the nations involved. Science diplomacy is the key to address all such international issues as it ensures the use of scientific collaboration among nations to address common problems and to build constructive international partnerships.

Science diplomacy has become an umbrella term to describe the formal or informal technical, research-based, academic, or engineering exchanges or collaborations. Considering S&T diplomacy with a specific approach is an issue that the political system of every country needs to take into consideration and provide the appropriate ground for benefiting from its related opportunities for their foreign and domestic stakeholders.

I am very happy that the NAM S&T Centre has recognised the import of science and technology diplomacy for development considering NAM and other developing countries have their own agenda to tackle the challenges. For this reason I understand that the Centre organised two international workshops within a period of two years, for the first time in any developing country. The one in Tehran, Iran, was held from 13-16 May 2012 under the title: 'Science and Technology Diplomacy for Developing Countries'. The second one, at Manesar (Haryana) India from27-30 May 2014 was under the theme: 'Perspectives on Science and Technology Diplomacy for Sustainable Development in NAM and Other Developing Countries'. Both the seminars, I understand, were very successful and well attended with participation of 20 countries in each seminar.

The present publication of the NAM S&T Centre comprises 14 papers from 12 developing countries that reflect the efforts under way by their respective governments for promoting science and technology diplomacy.

I am confident that this book will be a valuable resource material for use by experts engaged in the studies of inter-relationships of S&T with the foreign policies and the role of S&T diplomacy in the growth of nations.

(Ambassador Dr. V.B. Soni)

# Preface

The objective of this book is to present a different perspective of S&T Diplomacy from the angel of developing countries and their sustainable development. It is a collection of a number of paper presented in the workshop plus some other papers which add to the breadth and depths of the vision with which that book provide the readers.

As its main contribution, this book is expected to enable readers to reach an integrated understanding of the two seemingly decoupled layers of the institutions and organizations; the policy of Science, Technology and Innovation for sustainable development and Science and Technology diplomacy. These two sets of policies have often taken to be divided phenomena which are originated from two separated policy communities; the foreign policy community and the development policy community. The book aims to contribute to the literature by offering a unique and comprehensive conception which encompasses both domestic and foreign sets of policies simultaneously.

This double edged vision constitutes different dimensions therefore the reader can grasp it from different aspects. It is the shared theme in the papers which are devoted to the study of the diverse experiences of the developing countries at the national level. On the other hand, it is also underlined in the papers of which the level of analysis is regional and international. And finally, it is the centre of attention in the two papers which confer this concept from the theoretical aspect.

Therefore, the book is divided into four Sections:

## Section I: National Experiences

The first paper "Science and Technology Diplomacy: Progress of the Engineering Education in Cambodia" is written by Cambodian author, Chansopheak SEANG from the Institute of technology of Cambodia (ITC). This paper highlights initiatives in Cambodia centered on science and technology policies to serve its development agenda; wherein science diplomacy is an integral aspect. As the paper proceeds we learn that Cambodia's national development priorities for S&T and find out that the country focuses on technology development and application to strengthen to small and medium enterprises. This includes upgrading of standards and product quality, links between industry and research and development sectors, promotion of technologies for their benefits and supply of skilled labor.

The second paper belongs to Iranian authors, Mr. Birang, Amirinia and Ahmadi from V.P. for Science and Technology. They attempt to show the recent efforts of the Iranian society in strengthening the national innovation system and the structure supporting the S&T diplomacy. The authors enumerate some initiatives including different memorandums of Understanding among universities. Another program is the deployments of scientific attachés as well as technological attachés are among the Islamic Republic of Iran's (I.R. Iran) plans to expand scientific and technological ties with foreign countries. To facilitate international collaboration with internal organizations in science and technology, Iran's Ministry of Foreign Affairs set an agenda. Within the framework of this agenda, Iran's science and technology diplomacy and roadmap are taking shape and accordingly the Office of Science and Technology Cooperation is being established in the Ministry of Foreign Affairs to support to the country's international scientific and technological cooperation.

The next paper is authored by Mr. Siva Kumar Solay Rajah From Malaysian Ministry of Science, Technology and Innovation under the title of "Leveraging on Science, Technology and Innovation STI policy by Enhancing Collaborative Diplomacy". It addresses the need to have a well designed S&T diplomacy in order to have a successful STI policy. The author maintains in detail that Malaysia has designed such a policy to highlight the advantages of sharing STI that would bring mutual benefits and not conflict. This country has also strived to identify a common ground that will be acceptable when people from different countries exchange ideas, information and findings with regard to STI. Kumar concludes the aim of this integrated S&T policy-diplomacy is to encourage the notion that diplomacy should assist in resolving problems and not doing otherwise and encourage young leaders of various nations to invoke diplomacy in forging friendship and fostering continued relationship.

The fourth paper under the title of S&T Diplomacy: Status and opportunities for the Republic of Mauritius by Madhou, Suddhoo and Gokulsing from Mauritius Research Council and Ministry of Foreign Affairs, Regional Integration and International Trade. It highlights possibilities for a Small Island Developing State like Mauritius to intensify strategic relationships and to influence S&T policies in other countries. It illustrates, at first, the institutional settings of the ministries carrying out the missions of S&T diplomacy, namely, the Ministry of Foreign Affairs, Regional Integration and International Trade (MoFARIT) and the Ministry of Tertiary Education, Science, Research and Technology (MoTESRT) and explain their responsibilities in details and then discusses the possibilities for Mauritius to use S&T to emerge as leverage and enhance relationships with other countries with similar interest, such as the possibility for Mauritius to emerge as a role model among Small Island Developing States through the Vision of the Ocean Economy and The

Maurice Ile Durable (MID) Project and The potential for Mauritius to intensify its S&T strategic partnerships with Africa

The fifth paper is on the Nepalese experience of science and technology diplomacy. It started with a historical recount of UN and Nepal interactions dated back as early as 1950 and reviews its current partnership with several international and regional specialized organization on the issues related with science and technology diplomacy. The paper then proceeds to address the formation of an updated structure for the governance of S&T diplomacy within the country such as Nepal Academy of Science and Technology (NAST), Institute of Foreign Affairs (IFA) and concludes with the missions of these organizations. It is written by Chiranjivi Regmi, Nepal Academy of Science and Technology.

The last paper is written by Mr. Clifford Mupeyiwa, the Principal Science and Technology Officer of Zimbabwean government, who seeks to present Zimbabwean institutional framework of S&T policy and Diplomacy. It also addresses historical recounts of some of general milestones of S&T diplomacy at the international level.

## Section II: Regional Cooperation and South-South Relations

This chapter commences with the paper by Indian authors like Dr. Ruckmani Arunachalam, Dr. Rita Gupta, and Mrs. Sadhana Relia, from International Multilateral and Regional Cooperation Division, Department of Science and Technology. It is entitled "Better diplomacy and better science for better development" and is written to present the Indian initiatives to enhance S&T diplomacy among South counties. It enumerates and describes some of these programs such as 'New Africa Initiatives in S&T', Training of developing country scientists in India, Hosting developing country scientists for specialized training at the scientific centres and Implementation of Research Training Fellowships for Developing Country Scientists (RTF-DCS). It also goes on to explain some of Indian financial and scientific contributions to sustain international organizations and to maintain the level and quality of their collaborations.

The topic of the contribution of S&T diplomacy for enhancing the technical and vocational education and training (TVET) in Nigeria is discussed by Dr. Aworanti, Olatunde Awotokun (PhD). He is the Registrar/Chief Executive, National Business and Technical Examinations Board of Nigeria. The paper is entitled "Enhancing Technical and Vocational Education through Science and Technology Diplomacy" and explains how apt negotiations have been made by Nigerian government with the western world in the area of providing support for technical and vocational education and training (TVET) as well as in industrial development. In pursuit of multilateral relations, many international organisations still embark on supporting and partnering with Nigerian government in knowledge and skills development areas. This paper is therefore designed to evaluate the roles of science and technology diplomacy in enhancing technical and vocational education and training in Nigeria.

The next paper is under titled " The Nigeria's Technical Aid Corps Scheme, A Model for Science and Technology Diplomacy in Developing Countries was written by Dr. Bolarinwa Olugbemi of Raw Materials Research and Development

Council from Nigerian government. It is a paper to introduce a model for science and technology diplomacy in developing Countries based on the deployment of scientific knowledge, products and experts. As part of her foreign policy the Nigerian government established the Technical Aid Corps (TAC) Scheme in 1987, as an alternative to direct financial aid for African, Caribbean and Pacific (ACP) countries. It was designed not only to provide manpower assistance in all fields of human endeavour but also to represent a practical demonstration of South-South co-operation.

## Section III: International Organizations and Networking

The third section focuses its attention on the specific field of study, oceanography, nanotechnology and laser technology to highlight the benefits of international knowledge networking and S&T diplomacy at the global level:

The first paper of the section is authored by Venugopalan Ittekkot from Bremen University, Germany. It is about international science and diplomacy in the area of ocean and seas. It discusses the challenges and opportunities the oceans and seas offer to humanities welfare and nature sustainability. It, then, addresses some knowledge gaps existing in some of developing counties to deal with this issues and proceeds to propose actions at national, regional and international levels. In some cases, regional investments, networking and sharing of facilities and infrastructure could be beneficial. NAM S&T Centre could bring together member countries within an Oceanographic Network, where countries can support each other in the conservation and use of oceans and seas under their national jurisdiction. This could be in the field of education and research, national and regional policy making or in the design and implementation of regional oceanographic programs:

The next paper discusses the status of nanoscience and nanotechnology in many developed and developing countries as well as within the groupings of several countries, for example, BRICS, ASEAN and SAARC, and cooperation mechanisms adopted by them for the promotion of nanotechnology to meet their individual requirements. It is written by Ms. Radhika Tandon from NAM S&T Center. This paper discussed the significance of the issue of nano-technology transfer in the discourse of science diplomacy. Providing the fact that the North is deeply interested in exploiting the markets available in the developing countries for their nano-products, it is interested to engage South into different levels of negotiations and on the other hand, there is some level of mistrust in the developing countries about having been able to acquire 'real' technology from the developed countries even by paying high price, Therefore, South is more interested in the South-South cooperation. All these cross-interests and desirability of promoting partnerships to fulfil their own agenda vis-á-vis other nations requires negotiation and engagement of the nations in science and technology diplomacy.

The third paper of the third section was penned by Dr. Ihsan Fathallah Rostam from Iraq. It is on a specialized center for scientific research and treatment with laser. It starts with highlighting on the importance of regional cooperation, particularly in the fast growing areas like laser and nanotechnology, undertaking the following tasks: Dissemination of scientific knowledge, Information documentation and exchange of experience among workers in the field of laser from all disciplines (medical, engineering and physical, Scientific relations with corresponding centers regionally, within the organization and globally and everything would take care of and develop competence, curriculum development for the preparing intermediate stages enriching them with appropriate amounts of information regarding laser through adding and the new and developing present.

## Section IV: Theoretical Frameworks

The 4th Section is about conceptualization and theoretical modeling of the integration of S&T policy and Diplomacy. This task is completed by presenting two different cases studies; Turkey and Iran.

The first paper of this chapter is Dr. Tahereh Miremadi's paper. It aims to build a theoretical model to explain the interactive nature of dynamism of domestic public policy and diplomacy in the domain of science and technology. Bridging two different frameworks of Advocacy Coalition Framework and Double Edge Diplomacy, the paper attempts to show how domestic controversial policy advocacy of S&T determine the alternation of a country's position at the international arena and how the factor of policy brokering stabilizes this position by solidifying the relation between domestic and international policy communities. The presented model is applied on the case study of nuclear energy in Iran.

The second paper is written by Turkish authors Dr. Siir Kilkis and Ms. Nesibe Yazici from The Scientific and Technological Research Council of Turkey (TÜBITAK). They presented Turkish Vision for Science, Technology, and Innovation and begins by providing an overview of the increasingly more mature and vibrant R&D, innovation, and entrepreneurship system of Turkey as the basis of advancing opportunities to increase international cooperation. Such an overview is based on a unique application of the "functional dynamics" approach in the literature to characterize the Turkish innovation ecosystem and its extension to cover international co-operations.

The book is recommended for students, engineers, policy researchers, S&T planners and technical staff involved in S&T Diplomacy and sustainable development.

*Dr. Tahereh Miremadi*

*Mr. Abdul Haseeb Arabzai*

*Mrs. Sadhana Relia*

# Introduction

Science as an instrument has often been used to attend to problems of mutual interest and build constructive bilateral, regional and multilateral partnerships between the nations in the areas of strategic relevance, technology transfer, intellectual property rights, trade and commerce etc. In today's world, the innovating procedures, development and transfer of emerging and advanced technologies necessitate inter-government cooperation and in international dealings among nations, science as a diplomatic tool helps in removing political barriers offering tangible benefits to the concerned parties. Science diplomacy aids in fostering international collaborations among scientists in nations, including the ones where official diplomatic relations might be limited or strained, by providing a platform for scientists to cooperate. The potential of science and technology is slowly gaining recognition and many developing countries have initiated actions in leveraging international cooperation for national needs and priorities through science diplomacy and making new investments in human resources and infrastructure to enhance their S&T capabilities.

The Centre for Science and Technology of the Non-Aligned and Other Developing Countries (NAM S&T Centre) had organised an international workshop on 'Science and Technology Diplomacy for Developing Countries' in May 2012 in Tehran, Iran, and as a follow up and recognising the role of S&T Diplomacy in facilitating bilateral, regional and multilateral cooperation, the Centre organised yet another international workshop, the second in the series, on 'Perspectives on Science and Technology Diplomacy for Sustainable Development in NAM and Other Developing Countries' at Manesar (Haryana), India during 27-30 May 2014. This event was attended by 36 experts, diplomats and researchers from 22 countries, including Afghanistan, Cambodia, Colombia, Egypt, Germany, India, Indonesia, Iran, Malaysia, Mauritius, Myanmar, Nepal, Nigeria, Pakistan, Sri Lanka, South Africa, Switzerland, Syria, Turkey, Venezuela, Zambia and Zimbabwe.

The present book edited by Dr. Tahereh Miremadi of Iran, Mr. Abdul Haseeb Arabzai of Afghanistan and Mrs. Sadhana Relia of India is a follow up of the Manesar Science Diplomacy workshop and comprises 14 scientific papers contributed by the experts from 11 countries. The papers in this book have been categorised in four sections, namely, National Experiences, Regional Cooperation and South-South Relations, International Organisations and Networking and Theoretical Frameworks.

I put on record my appreciation for the efforts put in by the three co-editors of this publication for technical editing of the manuscripts. I also acknowledge the valuable services rendered by the entire team of NAM S&T Centre and am particularly thankful to Dr. (Mrs.) Kavita Mehra, Mr. M. Bandyopadhyay, Ms. Radhika Tandon, Ms. Keerti Mishra and Mr. Pankaj Buttan in compiling and checking the manuscripts, liaising with the authors, cover page designing, proof reading, formatting and taking all the necessary actions in giving a shape to this volume.

I am sure this book will be useful to researchers, policy makers and government officials of the developing countries who are engaged in international science and technology cooperation and deal with diplomatic negotiations on S&T affairs on behalf of their countries.

*Prof. Dr. Arun P. Kulshreshtha*
*Director General,*
*NAM S&T Centre*

# Contents

# — Section I —
# National Experiences

*Chapter 1*

# Science and Technology Diplomacy: Progress of the Engineering Education in Cambodia

*Chansopheak SEANG*

*ITC, Phnom Penh, Cambodia*
*e-mail: seang@itc.edu.kh*

## ABSTRACT

Science diplomacy is the use of science interaction among nations to address common problems faced by humanity and build constructive, knowledge based international partnership.The present paper highlights initiatives in Cambodia centered on science and technology to serve its development agenda; wherein science diplomacy is an integral aspect. In Cambodia, the concept of S&T Diplomacy is apparently new However, a deeper analysis of the context reveals that diplomacy between Cambodia and its regional and international partners allows this country to evolve and implement a robust S&T and Development Strategy. Cambodia's national development priorities for S&T are agriculture, engineering and technology, and natural science. It would like to focus on technology development and application to strengthen to small and medium enterprises. This would cover more than 31.00 SME's in the manufacturing sector with a special focus on 4 pillars for technology transfer to these enterprises. This includes upgrades of standards and product quality, links between industry and research and development sectors, promotion of technologies for their benefit and supply of skilled labor. This is aligned with the recognition that engineering plays an important role in the development of the SME's. The Institute of technology of Cambodia (ITC) is the oldest well know institute in Cambodia that offers an engineering degree. Education at the Institute spans the undergraduate and graduate levels by integrating the research activities. ITC has to integrate S&T diplomacy in its initiatives and sustain to cater to Cambodia's needs of human capital specialized in engineering and science to serve its development agenda. Some evidences of development of a favorable work environment for science and technology are presented with the fond hope that a much higher level of attention will be paid to take Cambodia along higher trajectories of growth; mediated by science and technology.

*Keywords: Cambodia, Diplomacy, S&T, Engineering education.*

# 1. Introduction

The key to enable and ensure sustainable development is dynamic and perpetual transformation; passing through phases of survival and growth. Science diplomacy in particular, embedded within frameworks of political diplomacy is an important medium to fulfill this goal of development. (1). It is important to re-emphasize that 'Science diplomacy is the use of scientific interactions among nations to address the common problems facing humanity and to build constructive, knowledge based international partnership Processes that sustain development should duly integrate economic development opportunities and rise up to emerging/novel challenges. Five important drivers of such a continual transformation are technological change, global production systems, changing labor markets, planetary boundaries (environment change), and demographic transition including migration) [2]. s[1].

The concept of science diplomacy is gaining increasing currency in the US, UK, Japan and other countries across the world. While the concept as such is still nebulous, it can usefully be applied to the role of science, technology and innovation in three related areas [1]:

☆ Informing foreign policy objectives with scientific advice (science in diplomacy);

☆ Facilitating international science cooperation (diplomacy for science) and

☆ Using science cooperation to improve international relations between countries (science for diplomacy).

# 2. Science and Technology in ASEAN and Cambodia

Science, technology and innovation can be powerful determinants and enablers of economic development, educational programs and protection of the environment. This view is shared by leaders in the ASEAN. Presently S&T cooperation in the ASEAN focuses on nine programme areas. These include (i) food science and technology (ii) biotechnology, (iii) meteorology and geophysics, (iv) marine science and technology, (v) non-conventional energy research, (vi) microelectronics and information technology, (vii) material science and technology, (viii) space technology and applications, and (ix) S&T infrastructure and resources development. A Sub-Committee coordinates and implements activities in each of these areas. Designated Dialogue Partners and Sectorial Dialogue Partners carry this agenda forward through mutually beneficial engagement. Several permanent bodies have been established with ASEAN Dialogue Partners such as China, India, EU and Russia, and with the Plus Three countries. Regular meetings among S&T officials from

ASEAN and Dialogue Partners are convened to promote closer cooperation. In addition, ASEAN also continues to closely collaborate in S&T with such international organizations as UNESCO and the WMO [3].

Cambodia 'is a member of the ASEAN since 1999. It has accordingly used diplomacy approaches to work closely with closely with ASEAN and other partners to improve the structure and policy dynamics of Cambodia's ministries/organizations that focus on S&T.

### 2.1. Infrastructure for S&T in Cambodia

Cambodia has a Council of Minister, that prepares the National Strategy Development Plan (NSDP). The National Committee on Science and technology (NCOST), created in 1999 coordinates and facilitates activities and tasks related to ASEAN initiatives in S&T. Cambodia's S&T affairs are handled by the ministries of Industry and Handicrafts (MIH),Mine and Energy (MME),Education Youth and Sport (MOEYS), Agriculture, Forestry and Fisheries (MAFF), Health (MOH), Public Works and Transport (MPWT), Post and Telecommunication (MPTC), Land Management, Urban Planning and Construction, Environment (MOE), Labor and Vocational Training, Commerce (MOC) and Economy and Finance (MEF)[4].

Importantly despite the fact that Cambodia has joined the S&T Cooperation of ASEAN, it has not developed comprehensive mid- and long-term plans and legal systems for science and technology development. Recently, the Ministry of Planning (MOP), supported by the Republic of Korea, is developing Cambodia's National Science and Technology Master Plan, 2014-2020, focusing on Establishing S&T foundation environment, Activating S&T climate, Securing R&D capabilities, and Empowering of S&T based core industry capabilities [5].

### 2.2. S&T in Higher Education of the Ministry of Education Youth and Sport (MOEYS)

As a member of COST the MOEYS has joined different activities related to S&T. It has its own structure, policy and strategy to achieve goals related to S&T.

The Master Plan supports Policy on Research Development in the Education Sector; as approved by the Ministerial Meeting on March 14th, 2011. The implementation of the Master Plan was financially and technically supported through the World Bank (WB) Higher Education Quality and Capacity Improvement Project (HEQCIP,2011-2015). The focus of the Policy on Research Development in the Education Sector is to enhance the quality of education, increase new knowledge and develop the society, economy and culture [6].

## 3. Engineering Education, Institute of Technology of Cambodia (ITC)

ITC is a higher education public institute under the MOEYS that focuses on engineering education. Diplomacy has a long history with this institute since its foundation in 1964, supported by cooperation between Cambodia and the former Soviet Union. In 1993, the governments of Cambodia and France agreed to renovate

ITC with a view to improve performance of administration and financial services along with the educational system of the institution and human resources.

ITC is presently at a major development cusp in South Eastern Asia where several partners converge. These include the French Coopération, Agence Universitaire de la Francophonie (AUF), La communauté Française de Belgique (CUD), AUN/SEED-net, GMSARN, and the School of Internet network. Through the cooperation, The Institute receives support, in terms of training and facilities, from a number of countries including France, Belgium and Japan. That allows ITC to develop the management structure and staff's capacity, the education systems and recently the research activities[7].

## 3.1. Improvement Management Structure and Staff of ITC

Based on a cooperation paradigm with universities in Europe and international organization, the structure of management of ITC involves three boards of management: These are the Board of council administration (Board of Trustees), Board of consortium, Board of directors. Every decision relating to policy, strategy, programs, budget, function of ITC has to pass through these boards.

All lecturers and staff engaged in research 40 PhD 117 M.Sc. and 76 Eng personnel, have an increasing set of opportunities to continue Master or PhD Program and receive training on specific skills (Figure 1.1). Country wise affiliations are also depicted for ready comparison.

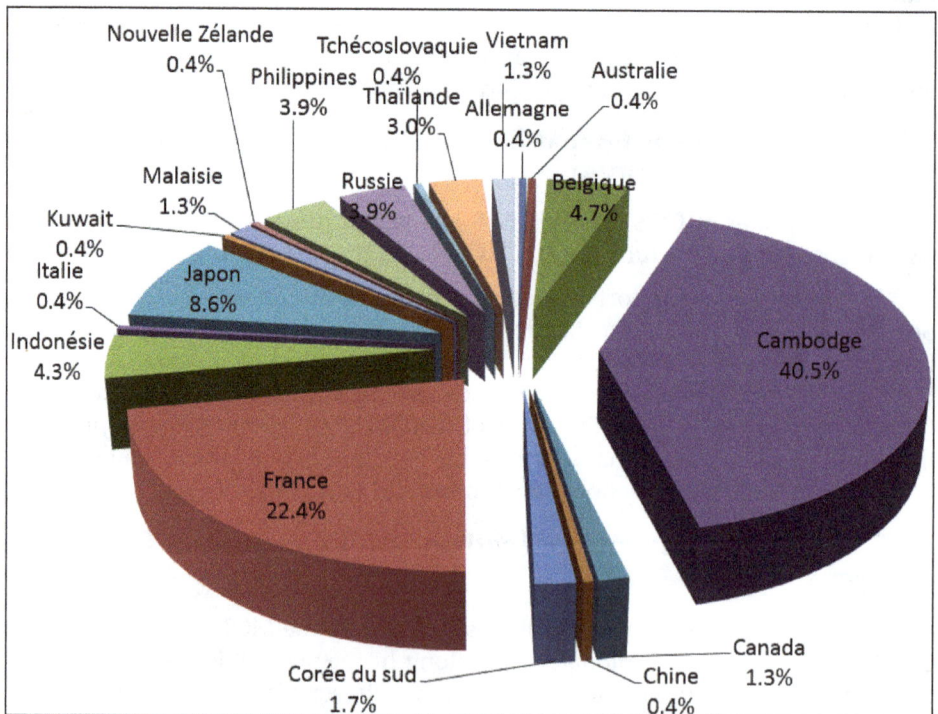

**Figure 1.1: Teaching Staff.**

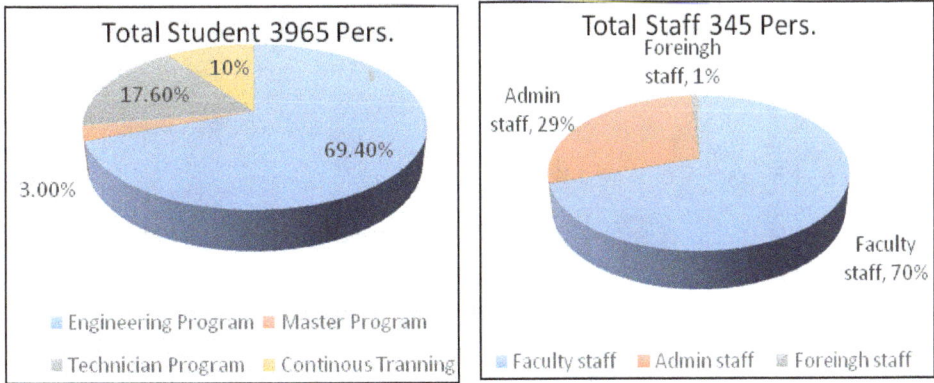

Figure 1.2: Staff and Student at ITC.

### 3.2. Engineering Programs

ITC elaborated new educational programs in accordance with the socio-economic reality of the country and the region.

### 3.3. Improvement of the Programs

It was started from the engineering degree to continuous programs, master programs and in 2015, the PhD Programs.

The engineering programs, that commenced with only five departments (civil engineering, food and chemistry, rural engineering, mine and industry and electrical and energy engineering)are improving with the inclusion of information and communication and geo-resource and geology.

### 3.4. The Continuous Programs

was launched in 2013 as three years programs, and now around 180 students are enrolled in some of the areas stated.

### 3.5. Graduate Programs

The master by research, 2 years programs, was launched in 2010. Currently 87 students enrolled in six masters programs in 2013-2014. The development of graduate programs get financial support from our partners (AUF, CUD). Exchange of teaching staff and research activities are with the cooperation of universities in France and Belgium. In the future we hope to get cooperation from universities in the ASEAN region.The PhD Sandwich Programs will be launch in 2015 with double degree with university in Belgium.

### 3.6. Research and Development

Of course, the engineering or sciences cannot succeed without related research activities. From 2003 to 2014, 56 projects were initiated and are in progress. This year 2013-2014 saw 28 projects (23 news), Figure 1.3,with total budget of 920.000USD. This was funded through international cooperation involving as the CUD (Belgium),

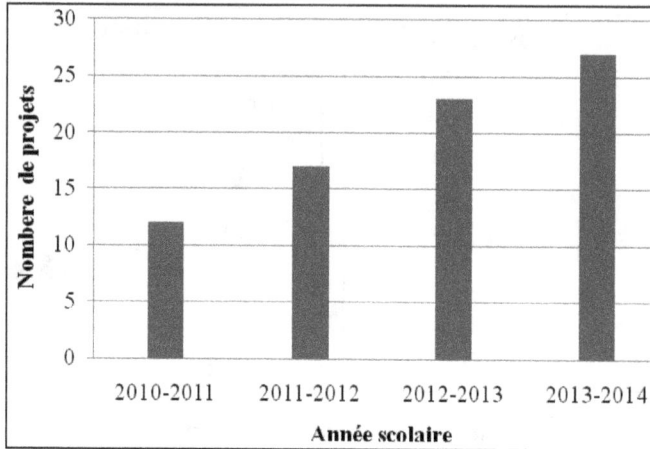

**Figure 1.3: Number of Research Project.**

AUF, AUN/SEED-Net JICA (Japan), Kanazawa University (Japan), Chugok EPCO Company (Japan), MOEYS-World Bank, The Asia Foundation, CGIAR Challenge Programme for water and Food and ICEM (Australia), Open Institute (Spain), Arup Sengupta. Foundation (Etats Unis), Chungnam National University (Korea), Takahashi Foundation (Japan), ODA-UNESCO (Japan).

These research projects are focused on water and waste water treatment, air quality, water quality, management of solid waste, water resources management and solid waste management, language processing, drying, production of energy from biomass, rural electrification, pyco-hydro, robot, treatment of arsenic in water.

### 3.7. Scholarship

ITC has been worked closely with AUF, French Embassy and a partner University in ASEAN. The number of scholarships are increasing over 2002 to 2013 as showed in the Figure 1.4. That allows the institute to give scholarship to students from Viet Nam and Laos to pursue engineering and Master Programs at the ITC. Currently 7 students from Laos national university (Lao)and 6 from the University of TraVinh (Viet Nam) are studying in different departments at ITC.

## 4. Challenge of S&T in Cambodia and ITC

☆ MoEYS receive only 1.8 per cent of GDP. HE received approximately 0.1 per cent of GDP.

☆ Funding for R&D is limited.

☆ Student's enrolment is: not consistent with priority programs. (less students register in S&T program; lack of access to right information that could help them make right decision [8].

☆ Awareness of S&T through the government and across all strata of the community is low.

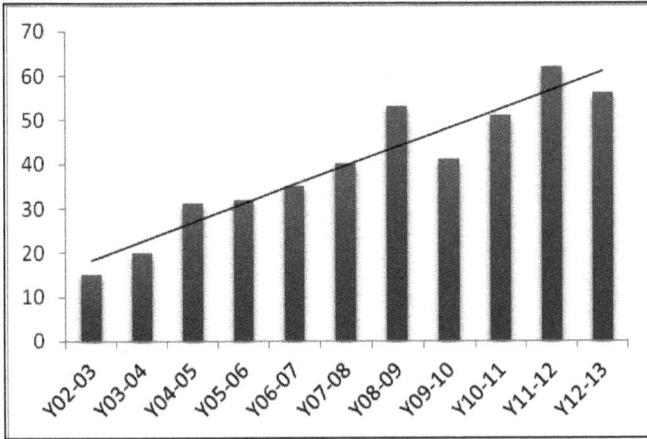

**Figure 1.4: Number of Student Received Scholarship to Study Abroad.**

Based on the above it is clear that Cambodia has a long way to go before it onlidates its own core strengths. An integrated science and technology policy that creates and embellishes human resources with competencies that establish the milieu for growth is needed. The present paper helps us with a preliminary snapshot of the learning and development milieu in Cambodia. A deeper analysis of the enabling environment and constraints that limit evolution is needed to set the context for policies.

## REFERENCES

1. New Frontiers in Science Diplomacy, Navigation the changing balance of power, ISBN 978-0-85403-811-4, The Royal Society, January 2010.

2. Sustainable Development solutions Network, A global initiative for the united nation, www.unsdsn.org/resources.

3. Communities Overview, Association of South-East Asian Nations. http://www.asean.org/communities/asean-socio-cultural-community/category/overview-36

4. Country report of Cambodia. Page 16-19.

5. Monthly Archives April 2013, Cambodia's National Science and Technology (S&T) Master Plan, Ministry of Planning, Cambodia http://snt.gov.kh/?m=201304 and lang=en.

6. Master Plan for Research Development in the Education Sector 2011-2015, Ministry of education, youth and sport. Cambodia. Approved at Ministerial Meeting on 14th March 2011.

7. biland'activités, réunion du consortium international d'appui, institut de technologie du Cambodge, Phnom Penh, les 26 - 27 mars 2014.

8. Policy on Cambodian Higher Education: High Education Vision 2030, Malaysia-Cambodia Joint Working Group Meeting on Higher Education, 23 Mai 2014, Phnom Penh Cambodia.

*Chapter 2*

# Science and Technology Diplomacy: Iran and the Path to Development

*Ali M. Birang[1], Hamid R. Amirinia[2] and Hossein Ahmadi[3]*

*[1]Deputy of International Affairs and Technology Exchange,*
*Vice- Presidency for Science and Technology,*
*The Islamic Republic of Iran,*
*e-mail: birang@isti.ir*
*[2]Advisor to the Vice-President for Science and Technology,*
*The Islamic Republic of Iran*
*e-mail: amirinia@citc.ir*
*[3]Manager of International Affairs,*
*Center for Innovation and Technology Cooperation,*
*The Islamic Republic of Iran*
*e-mail: h.ahmadi@citc.ir*

## ABSTRACT

In today's world, science and technology play a key role in providing various solutions for many of the problems we face as a society. They are accordingly important tools for sustainable development. However, such enablers as essentials, innovating procedures, development and transfer of emerging and advanced technologies have been exposed to some dramatic challenges. These have to be overcome only through cooperation amongst international science and technology networks. In this regard, the importance of exploring and opening new markets for economic production based on these technologies should not be neglected. Accordingly considering the great emphasis on promoting cooperation with most countries declared in important national documents such as the National Vision Document 1404 (2025), five-year development plans and creation of documents related to development of various technologies, Iran has increased bilateral and multilateral cooperation to strengthen the power of science and technology in the country and enhance the development of science and technology in other countries. The I.R. Iran demands engagement with a greater spread and depth of related processes in international science and technology collaboration. This should be especially through fair approaches and views that emphasize science and technology diplomacy for improved prosperity and not to oppress societies.

*Keywords:  Science and technology diplomacy, Development, Iran, Developing countries, International collaboration.*

## Introduction

Human endeavour ever since has always tried to derive the best from time, efforts and material inputs. This is seen even n science and technology cooperation amongst communities. However institutional forms of such an engagement have changed significantly as a result of expanding communication infrastructures.

Importantly, science and technology also empower and influence development, progress, and wealth generation. Promoting security and welfare through S&T, is a newly emerged concept in the 20th century, and well-established.

The pathway leading from scientific innovation to wealth creation has various links whose dynamic relation is the primary condition to achieve the desired result. As international interactions are increasingly developing in the current age, these links won't be necessarily geography-oriented; knowledge, expertise, capital, and market go beyond geographic borders and shape plenty of networks.

Recently, S&T diplomacy has emerged in the major hubs of technology and developed countries and due to its multilateral nature, different dimensions of this issue are still under consideration. The present moment in this evolution can also be seen as the stage of creating related literature on the subject in the global arena.

Development is a general concept, yet one of the factors of development of every country pertaining to the level of its scientific and technological progress. S&T with the highest value added is among the most powerful engines of development. Science diplomacy is an important element along the S&T pathway as a whole to prompt wealth generation and value. Any lapse in this tool in developing and less developed countries will be perceived as a missed link in this chain. Development of S&T is not absolutely domestic to grow entirely within a country but it also depends on international and foreign interactions. Such mutual interactions, either as importer or exporter of S&T can be promoted in a context of diplomacy and through application of S&T diplomacy which in turn will lead to different stages of S&T development.

## Materials and Methods

Iran has increased its bi- and multilateral cooperation in recent years through an understanding of the significance of S&T diplomacy in the current time. This will also strengthen the power of science and technology across bi – and multilaterals,. The writers thus attempt to present a brief introduction to Iran's efforts and measures for establishing S&T diplomacy. The present paper is a collaborative work which is result of library study of the available literature on the subject of S&T diplomacy in Iran. The first section; S&T Diplomacy in Developing Countries, provides a short review of the significance and status of the subject in the developing countries. The second part gives a brief account of Iran's achievements in S&T in recent years and Iran's measures and strategies for development of S&T through international cooperation and S&T diplomacy.

## S&T Diplomacy in Developing Countries

In today's world, science and technology is of paramount significance in providing various solutions to many of the existing problems and is considered as the only path towards sustainable development. By paying more attention to the role of newly emerging and advanced technologies and by considering their great economic and social benefits, countries have no alternative but to pay extra attention to such technologies. However, essentials, innovating procedures, development and transfer of emerging and advanced technologies have been subjected to some dramatic changes by shifting the paradigm towards such technologies. Due to the importance of tacit knowledge in advanced technologies previous concepts about transfer and development of such technologies have been altered. Development of these technologies, in addition to the necessity of cooperation and effective partnership among different major players in the field of science and technology, in the public and private sectors and their effective interactions, can only be realized through cooperation and playing an active role in different international science and technology networks. In this regard, the importance of exploring and opening new markets for economic production based on these technologies should not be neglected.

The new age of diplomacy with its especial features has provided multiple opportunities to foster interactions among the players of different areas. The environment of diplomacy and S&T has presented an extensive space for the participants of S&T and diplomacy and foreign politics at different levels all over the world. Accordingly, considering S&T diplomacy with a specific approach is an issue that the political system of every country should take into consideration and provide the appropriate ground for benefiting from its related opportunities for their foreign and domestic stakeholders.

Considering the above, it can be stated that developing countries should come out of their geographical shells and consider technological cooperation, new division of labour and the creation of new mass markets as part of their overall agenda to achieve development. Thus, the discussion of "science and technology diplomacy" becomes a necessity rather than a choice for all countries as it creates synergy between diplomacy and science and technology.

With regard to the countries' level of development (developing or developed), different levels can be considered for science and technology diplomacy in terms of national, regional and international aspects. However, what needs to be addressed primarily beyond the restricted ethnic, racial and geographical borders is the development of advanced technologies.

## S&T Diplomacy in Iran

Integration of historical teachings pertinent to learning and Islamic teachings to achieve prosperity through science, has led Iranians to be the origin of many achievements in this area as they have interacted and cooperated with other nations for many years. History of science in Iran, science-seeking intentions, efficient human resource, adequate S&T infrastructures, domestic and international Iranian technological achievements are factors paving the way for establishment of S&T diplomacy in Iran.

Yet, for many years, several internal and external factors have weakened this process preventing Iran from reaching its desirable, appropriate and effective position in science and technology. However, many Iranians, as global citizens are aiming to contribute to the development of science and technology. The active participation of Iranian elites in different Olympiads and their involvement in different fields of advanced technologies such as biotechnology, nanotechnology and aerospace are among the outstanding results of the mentioned endeavours.

Following the Islamic revolution and 8 years of imposed war, despite external pressures, Iran has executed many actions to promote science and technology status. Providing educational and professional training for the manpower as well as establishing mutual collaboration with scientific and research institutes in many countries can be named as examples.

Initially, the structure of science and technology in Iran was based on the linear model of science and technology evolution. In fact, the approach was to merely provide knowledge to strengthen and develop universities in the country. Therefore, beside state universities, non-state ones were established to train a remarkable number of researchers and students capable of knowledge creation. Thus, in 2011, the world witnessed Iran's fastest growth rate of science in the world. The number of ISI articles of Iran during 1970-2012 increased dramatically (Figure 2.1[1]) and as illustrated in Figure 2.2[2], with a growth rate of 20 per cent in the number of articles, Iran had the fastest growth rate among all nations in 2011, followed by China, South Korea, Spain, United Kingdom and United States. Likewise, Iran's rank in nanotechnology had moved up from 59th to 8th in the past decade (Figure 2.3).

The promotion of Market-pull approach led facilitator structures such as scientific and industrial towns, venture capital (VC) funds, science and technology parks, etc. to be developed in the country. As the result of the combination of the two approaches, Iran is benefitting from more than 500 research centers, nearly 200 universities, more than 150 incubators and science and technology parks and over 4 million students preparing the country for its scientific and technological leap.

In this regard and in line with the realization of knowledge-based society, preparing different plans and national documents for the development of different technologies such as country's Scientific Comprehensive Road Map is in progress.

---

1  http://www.nsf.gov/statistics/seind12/pdf/seind12.pdf.

2  NATURE VOL 480 22/29 DECEMBER 2011.

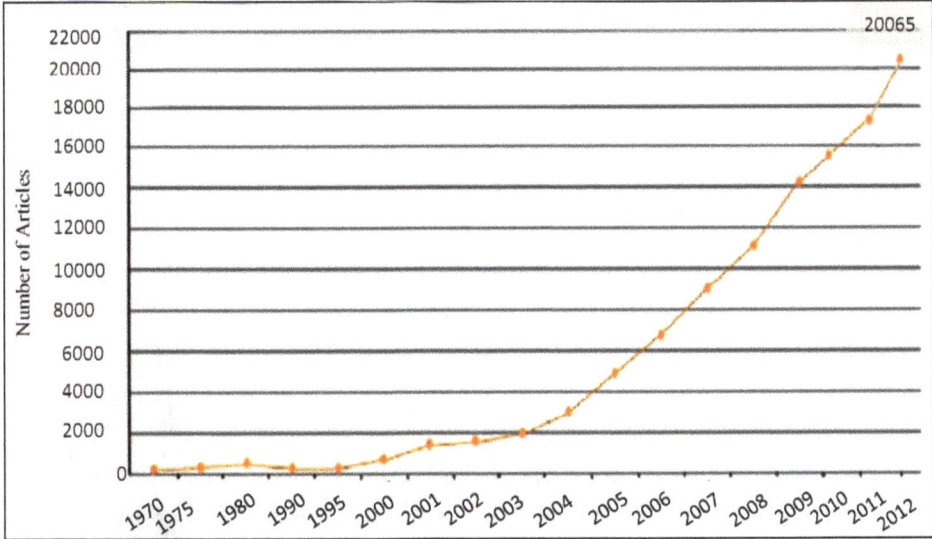

Figure 2.1: Growth of ISI articles of the Islamic Republic of Iran, 1970-2012.

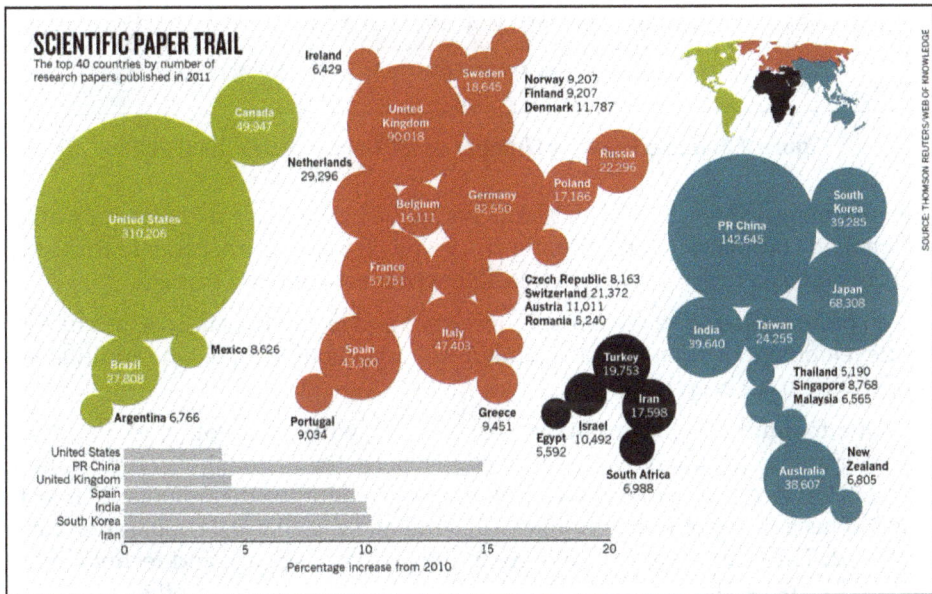

Figure 2.2: The Top 40 Countries by Number of Research Papers Published (2011).

It considers the organic connections between science and technology systems on one hand and economic, industrial, political and administrative environment on the other hand. Therefore, besides meeting national needs, country benefits from existing advantages of national and global technological priorities and global knowledge areas and creates mutual opportunities. With regard to emphasis of I. R.

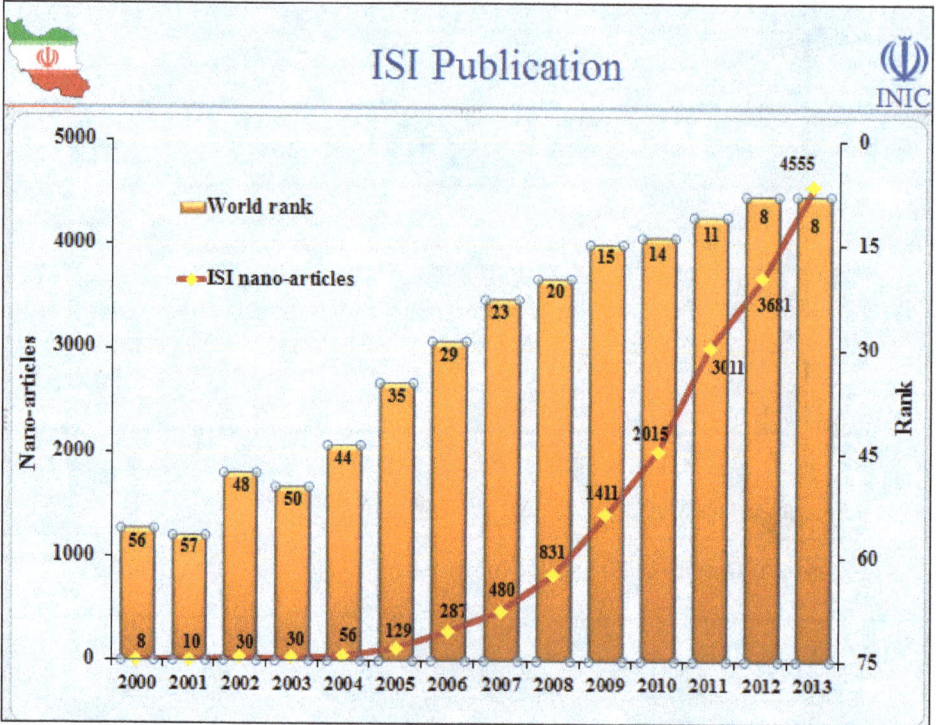

**Figure 2.3: Iran's Rank by Number of ISI Nano-articles 2000-2013.**

Iran's Supreme Leader, Ayatollah Khamenei on the formation of national innovation system and completion of its missing pieces, paying special attention to this issue will accelerate country's speed in fulfillment of the mentioned process.

In order to strengthen the power of science and technology in Iran and further the development of science and technology in other countries, bi- and multilateral cooperation have been increased.

Due to emphasis on promoting cooperation with most countries indicated in important national development documents, *i.e.* the national vision document 1404 (2025), five-year development plans and creation of documents related to development of various technologies, bilateral and multilateral cooperation have increased. These are meant to strengthen the power of science and technology in Iran and increase the development of science and technology in other countries. In line with playing role in global science field, ties with overseas counterparts have been developed in both state and non-state universities in Iran. For instance, more than 170 Memorandums of Understanding have been signed between Tehran State University and its 53 overseas counterparts from 1990 till 2009[3]. In almost all universities, Commercialization Offices have been launched to begin and maintain relationship between universities, market and industrial centers. Likewise,

---

3 http://international.ut.ac.ir/more.aspx?ID=30.

International Cooperation Offices have been established to adopt an appropriate approach towards international joint research and educational activities.

Deployments of scientific attachés as well as technological attachés are among the Islamic Republic of Iran's (I.R. Iran) plans to expand scientific and technological ties with foreign countries.

As a developing country and a member of the Non Aligned Movement (NAM), Iran viewed cooperation in science and technology as one of its top priorities and in 2012 hosted the first international workshop on Science and Technology Diplomacy for Developing Countries jointly with the NAM Science and Technology Centre in Tehran during which the representative members had the chance to exchange their views (Figure 2.4).

**Figure 2.4: First International Workshop on S&T Diplomacy for Developing Countries 13-15 May 2012.**

To facilitate international collaboration with internal organizations in science and technology, Iran's Ministry of Foreign Affairs set an agenda. Within the framework of this agenda, Iran's science and technology diplomacy and roadmap are taking shape and accordingly the Office of Science and Technology Cooperation is being established in the Ministry of Foreign Affairs to support to the country's international scientific and technological cooperation.

One issue to be addressed in Science and Technology Diplomacy is the policies adopted internally as well as externally. Applying a win-win policy that focuses on interaction and collaboration, a country can promote the synergy in actions of science and technology and accelerate the processes of development and prosperity. However, if the adopted policy is based on exploitation of others' resources, including human resources, especially the elites, the overall result will not be balanced and thus the cooperation in science and technology will develop to have a political nature. Today, in many countries, a large number of evidences indicate that many of the common problems of less developed and developing countries are rooted in the expansionist policies taken by other countries, and if this approach does not modify fairly, cooperation in science and technology will become corrupt, although it may seem reasonable.

Iran has made great attempts to convert sanctions' threats into opportunities and to develop science and technology through creativity and innovation. From another viewpoint, however, this situation has not only created some barriers to Iran's involvement in science and technology, it has also posed some restrictions ahead of the rest of the world too. The latter is a consequence of prevention of many Iranian talents to contribute to solving global problems.

The I.R. Iran demands to partake in international science and technology collaboration based on a fair approach. This is to emphasize the I R Iran's view that science and technology diplomacy as an approach should bring prosperity rather than oppression to the world. In this matter Iran's perspective is to implement a mutually reinforcing technology and diplomacy initiatives. Iran's novel achievements in medical drugs (such as the treatment of diabetic foot ulcers "Angipars") can be noted as a robust example in this regard.

## Some Important Overarching Insights and Conclusion

Promotion of scientific and technological capabilities in developing countries is among the most significant priorities of their long-term plans and up-stream documents. Hence, familiarizing the participants of this area - both diplomacy players and S&T players - with its significance should be put in the spotlight. Furthermore, to consider different dimensions of the subject, new potential roles of the two above-mentioned groups should be studied. Application of S&T diplomacy will be more possible and efficient only when this thought approximation is more practical and the gap between these players is lessened. Such common understanding among the players in developing countries can lead to common view on international and foreign interactions.

As S&T diplomacy is being systematized at the level of foreign politics, its role and status is being expanded, and domestic and foreign politics is being continued, there is no doubt that promotion of S&T will be accelerated in developing countries. As a result of international S&T relations in the context of S&T diplomacy, the current growth rate of developing countries will also be multiplied.

The gap between active elements of diplomacy and S&T is not just a mental one to be bridged by creating literature and convergence of thoughts. Lack of necessary mechanisms for efficient interaction of these two groups especially through synergistic approach is one of the greatest issues to come under consideration. There are other key subjects to be focused on - practical rules and executive guarantees, need for formation of intermediary elements, complementary role taking of public and private sectors, developing feedback mechanisms, and assessment of synergy to name a few.

Brain drain, lack of skilled human capital and scarcity of foreign investment in S&T are the consequences of insufficient convergence between diplomacy and S&T in developing countries. Heavy need for investment, market restrictions, limited access to international markets and non-optimal use of network structures available in the international S&T system further deteriorate the situation. Development of S&T diplomacy and its optimum application can reduce such negative outcomes.

By taking into consideration the aforementioned, there is hope that we will witness global efforts to realize prosperity in the world through the development of advanced technologies.

*Chapter 3*

# Leveraging on Science, Technology and Innovation (STI) Policy by Enhancing Collaborative Diplomacy

*Siva Kumar Solay Rajah*

*Ministry of Science, Technology and Innovation*
*Malaysia*
*e-mail: siva@mosti.gov.my*

## ABSTRACT

STI has been proven as an effective tool that can transform people's daily lives. It can be a significant and vital enabler to enhance a community's wellbeing. Hence, promotion of STI awareness should also include the field of research and technology development as well as the commercialisation of technologies.

A nation's STI policies is also a reflection of the nation's aspiration to tap STI's capabilities in transforming and addressing the challenges in various sectors such as health, education, agriculture, human capital and the community development. Thus, a comprehensive and transformative STI policy should include the role and the capabilities of STI in enabling engagement and strategic collaboration among various stakeholders.

While bureaucracy is required, it must, however, not hinder the creation of social and economic value to the nation. Building a good network of people's communication through STI is crucial for every nation especially more so for nation like Malaysia where strengthening existing relationship and developing new networks with partner countries at the regional and global levels are more important.

STI has significant influence on national, regional and global level hence in this context; STI can become an important tool and a strategic asset for diplomacy and the communication between people in various fields. STI also provides an avenue for countries to connect and collaborate to build a network for the mutual benefit of countries through right STI policies which also serves knowledge based collaboration between societies.

In short, STI policies must open doors and enable countries to capitalize on opportunities as well as ease constraints by building innovative societies. Malaysia has undertaken such efforts and the National Policy on Science, Technology and Innovation (NPSTI) 2013-2020 evolved to harness and leverage on STI.

*Keywords: Malaysia, Policy, Strategic, STI, Innovative, Wellbeing, Transformation, Economic.*

# 1. Introduction

The Ministry of Science, Technology and Innovation or MOSTI is the Malaysian ministry in charge of science, technology and innovation. It was created in 1973 by the federal government as the Ministry of Technology, research and Local Government and was reformed in 1976 as the Ministry of Science, Technology and Environment (MOSTE). Following the cabinet reshuffle of 2004, MOSTE evolved yet again to its current form. Today the Ministry is working on a number of initiatives to improve competitiveness in the field of science and technology through the generation of knowledge and sustainable development. MOSTI provides grants for research. Fund available to name a few are placed under specialized schemes, such as *Science fund, Technofund* (Pre-commercialization and IP acquisition fund) and *Innofund*.

STI is an integral part of any nation when transformation is needed. Malaysia's transformation journey is unique and challenging due to its unique social, economic and political scenario. The Malaysian government has taken bold steps in achieving a high-income nation status with an inclusive and sustainable economy by the year 2020.This journey has been outlined through the Economic Transformation Programme incorporating, among others, the 12 National Key Economic Areas and 8 Strategic Reform Initiatives (SRIs). These measures have been introduced to ensure that Malaysia achieves prosperous, sustainable and inclusive economy by the end of this decade.

In this context, the government of Malaysia launched The National Plan of Science, Technology and Innovation (NPSTI) which describes the agenda to make Malaysia more competent and competitive in STI. NPSTI provides an overview of the nation's required performance in STI which would guide the nation on strategies that have worked well which needs to be retained and charting revised and new strategies in areas that we have not done so well and areas that we need to be proficient.

Malaysia's ambition of moving ahead requires that we adopt bold and creative solutions. The business as usual approach will not work in an era fraught with uncertainties and intense global competition hence the general thrust of the NPSTI is to strengthen our basic foundations in the following areas:

&#9734; Our competency in generating and deploying knowledge through STI;

&#9734; Our STI human capital investment and development;

☆ Elevating the innovative capacity of our industry ;

☆ Strengthening our STI governance in place ;

☆ Developing a STI culture through engagement with our community.

This NPSTI initiative will help to create a better Malaysia which can meet the challenges of the 21$^{st}$ century. The document describes our nine-year agenda (2012-2020) to make Malaysia more competent and competitive in STI.

In a world of increasingly competitive and integrated global economy, empowering STI to play a vital role in the industrialisation and sustainable development of nations is crucial. The importance of these elements as important factors in the economic performance and social well-being of countries has become more evident in the face of globalisation, trade liberalisation and the emergence of knowledge-based industries. Indeed, countries that have witnessed impressive economic performance in recent years are those that have made heavy investments in STI.

## 2. A Policy Framework for a Better Future

Without sufficient mastery of STI, our rich natural endowments cannot be harnessed fully to propel the nation's economic trajectory to a higher plane. Our present inadequacies in STI are glaring and their manifestations are all too familiar. Our key industries are finding it increasingly difficult to ward off the challenges posed by our competitors who are closing down the gap on us through lower cost model and greater use of STI to achieve common objectives. We need to harness STI to move up the economic value chain. To do this, we must be prepared to confront our deficiencies in STI holistically and adopt a sustained effort to remedy our shortcomings. We understand that we must be prepared to abandon approaches that have not served us well and build upon those that have yielded positive results. Our policy document aims to clarify and achieve those objectives.

Much has transpired since the launch of the 2$^{nd}$ National Science and Technology (NSTP 2) in 2003. Many of the strategies under that policy are still valid. Some merely need to be tweaked to accommodate changing circumstances but for most new, bold and vigorous approaches still need to be adopted. These new approaches are necessary given the changed circumstances in the innovation landscape and expectations on public investments as follows:

### a. Changing Modes of Innovation

The innovation process has become a multi-institutional, multidisciplinary and a global endeavour. To succeed in an increasingly competitive global arena, Malaysia must strongly support partnerships, connectivity and inter-dependence that will enhance our knowledge and skills in order to respond to the changing modes of innovation.

### b. Harmonising the Various Sectoral STI Policies

Over the past three decades several policies and programmes have been introduced to advance the nation's efforts in boosting its STI capabilities. These include, among others, policies for biotechnology, green technology, nanotechnology and intellectual property rights. An overarching policy framework is crucial to harmonise our various STI initiatives to ensure that they are integral to the overall socio-economic thrust of the nation. Additionally, such a holistic framework aims to capture the synergies of the various initiatives, ensure coherence in our efforts and reduce duplications.

### c. Increasing the Role of STI System to Contribute to Socio-economic Well Being

Tighter budgets, affecting governments all over the world including Malaysia, are placing heavy demand for public investments such as STI to generate returns for the socio-economic well-being of the nation. Investments in science have a longer gestation period and these needs to be managed carefully to ensure that we derive maximum benefits from this investments. This challenge of budget must be overcome as STI will contribute greatly to the socio – economic well-being in the long run.

## 3. Moving Forward

In order to succeed and to face the future, Malaysia needs to innovate based on strong scientific fundamentals. In this regard, fostering strong and resilient partnerships, connectivity and inter-dependence amongst all sectors of the society is essential. For Malaysia, the changing landscape has become a challenge, not only to government but also to industries, universities, research institutes, and the whole STI ecosystem. Therefore, the formulation of National Policy on Science, Technology and Innovation (NPSTI) that adopts an integrated and holistic approach is timely to respond to these challenges. Our past approaches, though have yielded some results, have not had the desired effect on the development of our home grown capabilities in STI.

This overarching STI policy is also crucial to harmonise and consolidate all of our STI activities and programmes. It is indeed an essential component that should be placed at the centre stage of all national development plans and strategies.

## 4. Leveraging and NPSTI

Ideally, the NPSTI is focused on mobilising STI to steer Malaysia to become a high income nation that is sustainable and inclusive. The strategies and recommendations described in the document are aimed to laying out a framework that will guide our investments and initiatives that will generate positive outcomes for the nation. Building our strong foundations, we need to ensure that our efforts make a difference to people's lives. The new NPSTI describes an agenda to advance Malaysia towards a more competitive and competent nation built upon strong STI foundations. The policy is formulated based on the nation's achievements, challenges and lessons learnt. It charts new directions to guide the implementation of STI in creating a scientifically advanced nation for socio-economic transformation and inclusive growth.

The NPSTI is grounded on the following five fundamental foundations namely:

☆ STI for Policy;

☆ Policy for STI;

☆ Industry Commitment to STI;

☆ STI Governance; and

☆ STI for a stable, peaceful, prosperous, cohesive and resilient society.

To ensure success and achievement, the above five foundations should embody the following six strategic thrusts:

☆ Advancing scientific and social research, development and commercialisation;

☆ Developing, harnessing and intensifying talent;

☆ Energising industries;

☆ Transforming STI governance;

☆ Promoting and sensitising STI; and

☆ Enhancing strategic international alliances.

The NPSTI framework is depicted as follow:

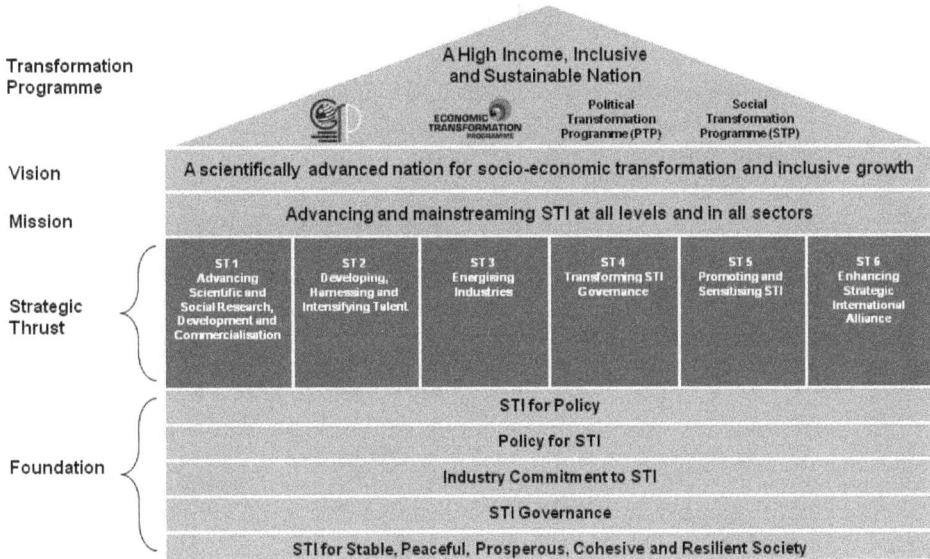

## 5. Enhancing Collaborative Efforts

We know a fact that the world economy is undertaking a rapid globalisation. This has, among other things, led to faster information flow, global alliances, global manufacturing, and talent mobility. Countries have no option anymore but to strengthen existing linkages and build new alliances. Malaysia should not be a bystander in this wide interconnectedness world but to collaborate, co-create and foster strategic partnerships for socio economic growth.

Malaysia is a member to various organizations and has participated in many forums at regional and international levels such as the Association of Southeast Asian Nations (ASEAN), the Asia-Pacific Economic Cooperation (APEC), Non-Aligned Movement (NAM), Organisation of Islamic Cooperation (OIC), the World Trade Organisation (WTO), the World Health Organisation (WHO) and United Nations Educational, Scientific and Cultural Organisation (UNESCO). In the past, Malaysia has also signed a number of Memorandum of Understanding (MoU) in the S&T cooperation with partner countries since 1968 and the Free Trade Agreement (FTA) in the bilateral and multilateral levels. In this regard, some of the past agreements and MoUs needed to be reviewed and revised to ensure optimal benefits to the country in terms of human capital development cooperation, technology transfer and trade opportunities.

It is therefore crucial for Malaysia to strengthen its existing relationships and develop new networks together with the partner countries at the regional and global levels. In 2011, the ASEAN market, for instance accounted for 26.2 per cent of the country's total trade for 2011. While emerging markets, like India, China and the countries of Eastern European Block, needed to be further explored and tapped.

Hence, by having a policy with strong fundamentals which encourages the enhancement for global linkages and partnerships, the following policy measures will be undertaken:

☆ Improve Research, Development and Commercialisation (R,D and C) ecosystem to attract global partners

☆ Nurture domestic talents to enable organisations and industries to penetrate global markets

☆ Develop partners, allies and channels in key destination countries

☆ Establish "go-global" market strategies for home grown STI product (including market access, business intelligence, etc.)

☆ Strengthen marketing and development of global brands

☆ Continuous improvement in monitoring and evaluation

☆ Intensify domestic and international networks for research collaboration, strategic partnerships and business relationships

So much so, encouraging international cooperation in STI is a crucial enabler for policy planning and to achieve a nation's objective for a sustainable economic growth in the face of global challenges. However, such initiatives should provide an avenue and ease of information flow and technology transfer that could build lasting relationship between countries. Malaysia currently has bilateral STI cooperation agreements with following countries; Australia, US, Brazil, the Republic of Korea, Egypt, Kenya, Hungary, New Zealand, Pakistan, Poland Tunisia, Russia, Vietnam, Uzbekistan, Iran and Syria. The activities in these agreements generally cover conducting joint projects of mutual interest, exchange of date and visits of scientists, holding of joint workshops and seminars as well as technical training.

Malaysia is also active in the Association of South East Nations (ASEAN) working and building constructive partnerships through ASEAN's Committee on Science and Technology with Indonesia, Brunei, Thailand, Singapore, Philippines, Myanmar, Lao PDR, Cambodia and Vietnam, championing issues relating to STI. Such cooperation will help Nation to bridge and strengthen relations between countries. While Malaysia is striving to improve the STI agenda, a lot need to be done. As a general overview, the Table 3.1 show Malaysia's standing in terms of competitiveness and R&D expenditure.

This table shows Malaysia's ranking in the World Competitiveness Yearbook (WCY) from 2009 to 2013. In 2013, Malaysia was ranked 15th down one notch from her previous rank (14th) in 2012. The Malaysian Productivity Corporation (MPC) is responsible to consolidate and provide the input to WCY, and MOSTI is one of the data contributors for

## 6. Banking on STI and Diplomacy

We may not be able to specify the definition of 'diplomacy' completely but one of the definitions that could be considered is "Diplomacy is a complex and often challenging practice of fostering relationships around the world in order to resolve issues and advance interests" (http://diplomacy.state.gov/discoverdiplomacy/). Intriguingly, the concept of fostering relationship is challenging due to the acceptability and the readiness of countries to share knowledge when it comes to STI. Surely, if a nation wanted to share their expertise in STI, it must have strategized their reason of doing so. Hence, when relating to fostering STI diplomacy the following perspective or aspects could be considered:

- ☆ Realization of the advantages of sharing STI that would bring mutual benefits and not conflict
- ☆ Identity a common ground that will be acceptable when people from different countries exchange ideas, information and findings with regard to STI
- ☆ Encourage the notion that diplomacy should assist in resolving problems and not doing otherwise
- ☆ Encourage young leaders of various nation to use diplomacy using STI elements in forging friendship and fostering continued relationship

The Figure 3.1 shows the comparison of gross expenditure on R&D (GERD) as a percentage of GDP (per cent) for selected countries. For fiscal year 2011, Malaysia was ranked at 35th position with GERD/GDP value at 1.07 per cent.

## 7. Banking on STI and Diplomacy

We may not be able to specify the definition of 'diplomacy' completely but one of the definitions that could be considered is "Diplomacy is a complex and often challenging practice of fostering relationships around the world in order to resolve issues and advance interests" (http://diplomacy.state.gov/discoverdiplomacy/). Intriguingly, the concept of fostering relationship is challenging due to the acceptability and the readiness of countries to share knowledge when it comes to STI.

**Table 3.1: IMD World Competitiveness Yearbook (WCY) Rankings by Countries, 2009-2013**

| Sl.No. | Country | 2013 Rank | 2013 Score | 2012 Rank | 2012 Score | 2011 Rank | 2011 Score | 2010 Rank | 2010 Score | 2009 Rank | 2009 Score |
|---|---|---|---|---|---|---|---|---|---|---|---|
| 1. | USA | 1 | 100.0 | 2 | 97.8 | 1 | 100.0 | 3 | 99.1 | 1 | 100.0 |
| 2. | Hong Kong | 3 | 92.8 | 1 | 100.0 | 1 | 100.0 | 2 | 99.4 | 2 | 98.1 |
| 3. | Singapore | 5 | 89.9 | 4 | 95.9 | 3 | 98.6 | 1 | 100.0 | 3 | 95.7 |
| 4. | Germany | 9 | 86.2 | 9 | 89.3 | 10 | 87.8 | 16 | 82.7 | 13 | 83.5 |
| 5. | Qatar | 10 | 85.5 | 10 | 88.5 | 8 | 90.2 | 15 | 83.8 | 14 | 82.0 |
| 6. | Taiwan | 11 | 85.2 | 7 | 90.0 | 6 | 92.0 | 8 | 90.4 | 23 | 75.4 |
| 7. | **Malaysia** | **15** | **83.1** | **14** | **84.2** | **16** | **84.1** | **10** | **87.2** | **18** | **77.2** |
| 8. | Australia | 16 | 80.5 | 15 | 83.2 | 9 | 89.3 | 5 | 92.2 | 7 | 88.9 |
| 9. | United Kingdom | 18 | 79.2 | 18 | 80.1 | 20 | 80.3 | 22 | 76.8 | 21 | 76.1 |
| 10 | China Mainland | 21 | 77.0 | 23 | 75.8 | 19 | 81.1 | 18 | 80.2 | 20 | 76.6 |
| 11. | Korea | 22 | 75.2 | 22 | 76.8 | 22 | 78.5 | 23 | 76.2 | 27 | 68.4 |
| 12. | Japan | 24 | 74.5 | 27 | 71.4 | 26 | 75.2 | 27 | 72.1 | 17 | 78.2 |
| 13. | Thailand | 27 | 73.0 | 30 | 69.0 | 27 | 74.9 | 26 | 73.2 | 26 | 70.8 |
| 14. | Philippine | 38 | 63.1 | 43 | 59.3 | 41 | 63.3 | 39 | 56.5 | 43 | 54.5 |
| 15. | Indonesia | 39 | 61.8 | 42 | 59.5 | 37 | 64.6 | 35 | 60.7 | 42 | 55.5 |
| 16. | India | 40 | 59.9 | 35 | 63.6 | 32 | 70.6 | 31 | 64.6 | 30 | 66.5 |

*Note*: Ranking over 60 countries (2013); 59 countries (2012); 59 countries (2011); 58 countries (2010); 57 countries 2009).

*Source*: IMD World Competitiveness Yearbook 2009-2013.

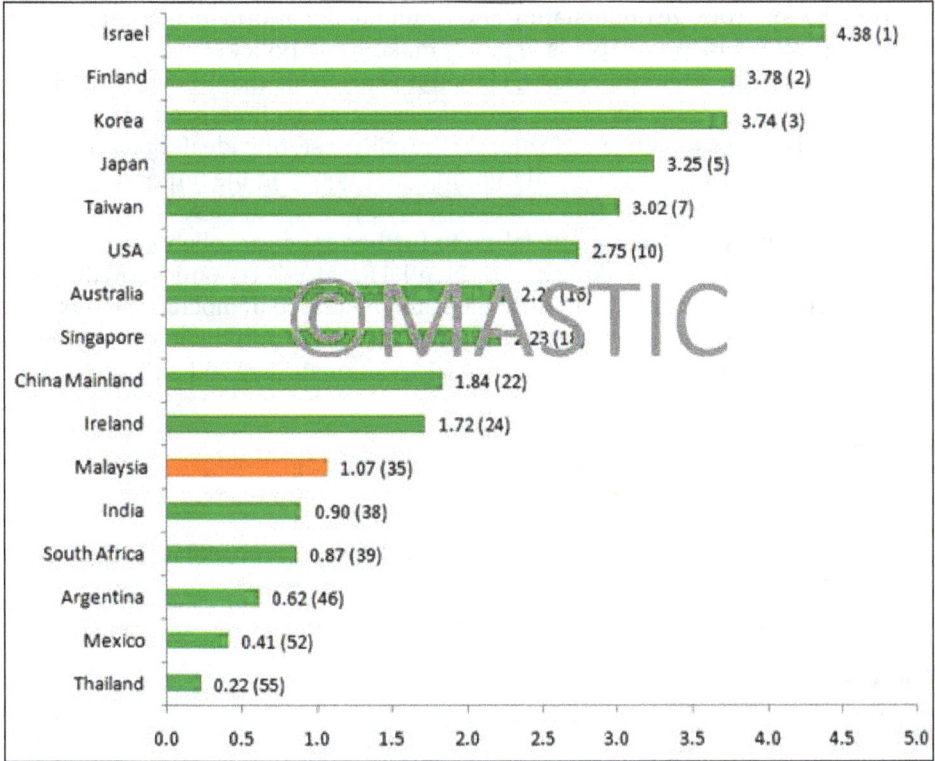

**Figure 3.1: Comparison R&D Expenditure and GDP (Per cent) by Countries.**

Surely, if a nation wanted to share their expertise in STI, it must have strategized their reason of doing so. Hence, when relating to fostering STI diplomacy the following perspective or aspects could be considered:

☆ Realization of the advantages of sharing STI that would bring mutual benefits and not conflict

☆ Identity a common ground that will be acceptable when people from different countries exchange ideas, information and findings with regard to STI

☆ Encourage the notion that diplomacy should assist in resolving problems and not doing otherwise

☆ Encourage young leaders of various nation to use diplomacy using STI elements in forging friendship and fostering continued relationship

## 8. Conclusion

It is not easy to formulate a policy without inserting the need to establish and enhance international cooperation. To encourage STI diplomacy, one must take into account that any collaboration must be backed by policy measure that will enable a nation to do so for the sake of progress. Thus, STI diplomacy could be an important way to bridge understating and foster ideal platform to help countries in becoming better and more capable. If policy makers and policy strategist formulate STI policy that can internationalise STI then it is probable that STI will contribute towards re-engineering economic growth, assist as well as promote regional stability and could encourage, STI values eventually fostering ideals of meritocracy, transparency, creative ideas, critical thinking, and boosting the importance of STI towards building a better nation. Ultimately, Malaysia has done so by formulating a comprehensive policy within its fabric to further enhance collaboration in STI. This initiative creates the thinking of the potential for rapid economic growth, faster expansion of the lower and the middle class, and increased democratic governance as well as increased trade between countries becoming an optimistic scenario. Such future scenarios, including some that are quite pessimistic, are laid out in the report Global Trends 2030.

## REFERENCES

### Publication

1.  National Plan on Science, Technology and Innovation Malaysia 2013-2020 (Unpublished Document)

2.  2. 10th Malaysia Plan (2011-2015), available for download at www.epu.gov.my/

### The Web

3.  http://www.isis.org.my/attachments/EN_AsiaPacBull_261dt08May2014.pdf

4.  http://www.mastic.gov.my/en/web/guest/wcy

5.  http://diplomacy.state.gov/

6.  http://globaltrends2030.files.wordpress.com/2012/11/global-trends-2030-november2012.pdf

*Chapter 4*

# S&T Diplomacy: Status and Opportunities for the Republic of Mauritius

*M. Madhou[1], A. Suddhoo[1] and D.P. Gokulsing[2]*

[1]*Mauritius Research Council,*
*Government of Mauritius*
*e-mail: m.madhou@mrc.intnet.mu*
[2]*Ministry of Foreign Affairs,*
*Regional Integration and International Trade,*
*Government of Mauritius*

## ABSTRACT

Science Diplomacy is a field which is gaining a lot of attention in Mauritius. This paper presents the main approaches in establishing international collaborations in Science and Technology and highlights possibilities for a Small Island Developing State like Mauritius to intensify strategic relationships and to influence S&T policies in other countries.

In Mauritius the responsibility of diplomatic relations with other countries falls under the Ministry of Foreign Affairs, Regional Integration and International Trade (MoFARIT). The responsibility of Science and Technology falls under the Ministry of Tertiary Education, Science, Research and Technology (MoTESRT). The MoTESRT also deals directly on regional and international cooperation on matters related to Tertiary Education, Science, Research and Technology. Besides the MoTESRT, other S&T related Ministries and government institutions are involved in a number of partnerships in their respective specific fields. The Science Diplomatic Efforts of Mauritius have culminated in ratifications of a number of science-related international treaties, international relationships for infrastructural S&T facilities. Even if Mauritius has been successful to some extent in using international science collaborations to address common problems and to build constructive, international partnerships, the opportunity to be propelled as S&T key player on the global scene has not been fully tapped by our current efforts. This paper discusses the possibilities for Mauritius to use S&T to emerge as leverage and enhance relationships with other countries with similar interest.

Three areas of opportunity are discussed:

1. The possibility for Mauritius to emerge as a role model among Small Island Developing States through the Vision of the Ocean Economy and
2. The Maurice Ile Durable (MID) Project
3. The potential for Mauritius to intensify its S&T strategic partnerships with Africa.

*Keywords: Science diplomacy, Mauritius, Ocean economy, Small island developing state.*

## 1. Introduction

The Republic of Mauritius is a Small Island Developing State (SIDS) situated in the Indian Ocean, approximately 2400 kilometers off the South East Coast of Africa. It covers a surface area of 2,040 square kilometers and includes the main island of Mauritius (1865 km²), the island of Rodrigues (108 km²), the two Agalega-Islands and the Cargados-Carajos-Archipelago. It was colonized successively by the Dutch French and British. The on-going global financial and economic crisis threatens Mauritius given its inherent vulnerabilities as a Small Island Developing State (SIDS). The island is dependent on a narrow range of income generating sectors and is vulnerable to adverse conditions linked to climate change and the energy crisis. These challenges, added to the present situation of globalization and trade liberalization have fuelled the need for Mauritius to reposition itself in the global economy. It is recognized that science, technology and innovation are crucial in addressing these challenges and many countries are investing significant efforts to upgrade human competencies in science, technology and innovation, strengthen S&T resources and accelerate research and development (R&D) and innovation.

The creation, for the first time in 2010, of a Ministry of Tertiary Education, Science, Research and Technology, shows the strong commitment of the Government to make tertiary education as well as S&T important pillars of the Mauritian Economy. The vision of the Ministry is to transform Mauritius into a Knowledge-Based Economy by 2022, a vision which requires the concerted actions of all stakeholders and which calls for a revision of our policies and strategies to sharpen our competitive advantage and to increase our resilience to global threats.

It is widely acknowledged that the majority of challenges facing society are of global concern; and science, technology and their applications can address these challenges. As a result science and technology are becoming increasingly embedded into international relations of countries. At the same time, it is recognized that science diplomacy can help to improve relations between countries, involving a strategic link between Science and Diplomacy which gives another dimension to the S&T system.

The Republic of Mauritius has been actively involved in a number of regional/ international S&T initiatives. This paper is an attempt to identify firstly the main approaches and diplomatic efforts' used by Government institutions to establish international S&T cooperation programs, secondly to identify existing programs that can enhance the science diplomatic efforts of a SIDS like Mauritius to influence foreign S&T policy objectives and thirdly to identify common areas of concern which require concerted action among countries in the region.

## 2. Methodology

This study is a desktop study and involved the following activities:

1. A brief situational analysis of the main approaches used by Government Bodies to establish international collaborations in Science and Technology.

2. Identification of programs which required Science Diplomatic Efforts. In this line the joint effort of the Republics of Mauritius and Seychelles for the joint jurisdiction of about 396000 km$^2$ for the Extended Continental Shelf in the Mascarene Plateau region will be used as an example.

3. S&T programs which have been initiated in Mauritius and which could influence foreign S&T policy objectives and activities.

4. Identification of S&T programs which could intensify the strategic relationships of Mauritius in the region.

## 3. Results and Discussion

### The Present Landscape

In Mauritius the responsibility of diplomatic relations with other countries falls under the Ministry of Foreign Affairs, Regional Integration and International Trade (MoFARIT). The responsibility of Science and Technology falls under the Ministry of Tertiary Education, Science, Research and Technology (MoTESRT). The MoTESRT also deals directly on regional and international cooperation on matters related to Tertiary Education, Science, Research and Technology. Besides the MoTESRT, other S&T related Ministries and government Institutions are involved in a number of technical/scientific partnerships (*e.g.* Technical Programs, MoUs, Regional Projects) in their respective specific fields.

S&T partnerships with other countries include:

1. MoUs and Agreements
2. Participation in Regional and international projects
3. Signatory to international conventions
4. Membership/Affiliations to regional and international instances

It is noteworthy that most of the partnerships that the country is involved are 'Technically driven', 'Policy related' or 'Science Driven'. Such partnerships can be grouped under the Diplomacy for Science initiatives and they have culminated in ratifications of a number of science-related international treaties, international relationships for infrastructural S&T facilities such as establishment of Rajiv Gandhi Science Centre, setting up of Radio Telescope facility in Mauritius by the Indian Government and more recently in November 2013 there has been the signature of a Memorandum of Understanding between the Mauritius Research Council (on behalf of the Ministry of Tertiary Education, Science Research and Technology) and the Indian Institute of Technology Delhi, India with respect to the development and implementation of an IIT like institution in Mauritius to provide a world class research based educational platform.

However one important aspect of Science Diplomacy is its application as a 'soft power' tool, to attract, persuade and influence building on common interests and values (Nye, 2004). This aspect, referred to as Science for Diplomacy, has been largely associated to powerful nations such as the U.S and Japan. This aspect of S&T diplomacy is not largely documented for Small Island Developing States like Mauritius. Furthermore there is an untapped potential for Mauritius to inform foreign policy objectives with scientific advice (Science in diplomacy)

Hence even if Mauritius has been successful to some extent in using international science collaborations to address common problems and to build constructive, international partnerships, the opportunity to be propelled as S&T key player on the global scene by using Science for Diplomacy and Science in diplomacy has not been fully tapped in our efforts. The following sections discuss the possibilities for Mauritius to use S&T to emerge as a key player and intensify strategic partnerships with other countries. Three areas of opportunity have been identified and will be discussed in this line.

## Areas of Opportunity to Intensify Strategic Partnerships with SIDS

The on-going global financial and economic crisis further threatens Mauritius given its inherent vulnerabilities as a Small Island Developing State (SIDS). The island is dependent on a narrow range of income generating sectors and is vulnerable to adverse conditions linked to climate change and the energy crisis. These challenges, added to the present situation of globalization and trade liberalization, have fuelled the need for Mauritius to reposition itself in the global economy. The active search for continuous improvement in this direction has led to the emergence of 2 new initiatives that could change the socio-economic trajectory of Mauritius-the Ocean Economy and the Maurice Ile Durable(MID) (Sustainable Mauritius) Vision. These initiatives also represent an opportunity for Mauritius to be a role model for other SIDS. Science and Technology are major elements required for the success of both initiatives and along this line S&T cooperation programs could eventually be a means to intensify the partnerships among SIDS.

### The Ocean Economy

The Republic of Mauritius is a SIDS situated in the Indian Ocean, approximately 2400 kilometers off the South East Coast of Africa. It covers a surface area of 2,040 square kilometers and includes the main island of Mauritius (1865 km$^2$), the island of Rodrigues (108 km$^2$), the two Agalega-Islands and the Cargados-Carajos-Archipelago. Mauritius has one of the largest Exclusive Economic Zones in the world. Moreover, in 2012, the United Nations approved the coordinates submitted jointly by Mauritius and Seychelles for jurisdiction over an area of our continental shelf extending over 400 000 km$^2$. Mauritius has now a total area of 2.3 million km$^2$ over which it can exercise its sovereignty.

The bilateral relations and diplomatic ties between Mauritius and Seychelles were further strengthened by the entry into force of the two Treaties concerning the joint exercise of sovereign rights by Seychelles and Mauritius over the Extended Continental Shelf in the Mascarene Plateau region and the joint management of the Extended Continental Shelf, which were signed in March 2012 during the State Visit of the President of Seychelles to Mauritius (Government Information Service, 2012). The potential now exists for Seychelles and Mauritius to develop mechanisms to explore the extended continental shelf and manage its resources. It is worth mentioning that two projects to tap the potentials in the ocean have been initiated in Mauritius. These are the exploitation of deep sea water for various applications such as bottled drinking water, air conditioning, aquaculture, thalassotherapy and the drafting of a long term strategy to turn Mauritius into a petroleum hub. Furthermore Mauritius has already developed a Road Map to the Ocean Economy and 7 clusters (Marine Biotechnology, Marine Renewable Energy, Seabed Exploration for Hydrocarbon and Minerals, Fishing/Seafood Processing/ Aquaculture, Deep Sea Water Applications, Marine Services, Seaport Related Activities and Ocean Knowledge) have been identified for further development (Prime Minister's Office, 2013). These initiatives offer an unprecedented scope for Mauritius to use S&T to further consolidate its diplomatic links with Seychelles but also with other SIDS by being a role model.

Ocean S&T cooperation programs would also be of interest to countries of Eastern and Southern Africa, in which case regional mechanisms to harness resources and expertise towards interlinked problems of the coastal and marine environment would be relevant. It is noted that Mauritius has ratified the Nairobi Convention for the Protection, Management and Development of the Marine and Coastal Environment of the Eastern African Region which came into force in 1996 and amended in 2010.

Science diplomacy networks can be most useful to the development of the Ocean Economy. However they should be designed to address common issues such as piracy, bio-prospecting, maritime surveillance and security. There is a clear opportunity for Mauritius to take a leading position in these fora. For this to be successful there is a need for Mauritius to integrate science with traditional diplomacy, this close partnership between scientists and diplomats can also help to brand the image of the island as an innovative nation.

## Sustainable Development: The Maurice Ile Durable Concept

### Sustainable Development and International Partnerships

The Ministry responsible for sustainable development in Mauritius is the Ministry of Environment and Sustainable Development (MoESD). This Ministry is already implementing the recommendations of major international sustainable development summits in collaboration with other Ministries; these include recommendations from the UN Conference on Environment and Development-Rio De Janeiro, 1992, World Summit on Sustainable Development, Johannesburg, 2002, UN Conference on Sustainable Development, Rio De Janeiro, 2012, Barbados Programme of Action,1994, Mauritius Strategy for Implementation of the BPoA,

2005. Furthermore the MOESD is implementing Multilateral Environmental Agreements such as the UN Framework Convention on Climate Change, Montreal Protocol on Substances that Deplete the Ozone Layer, the Stockholm Convention on Persistent Organic Pollutants, Miniamata Convention on Mercury and Convention on the Protection, Management and Development of the Marine and Coastal Environment of the Eastern African Region and related protocols. The MoESD is also a member to international organizations such as Indian Ocean Commission (IOC), Common Market for Eastern and Southern Africa (COMESA) and the Southern Africa Development Community (SADC). Some international partnerships in terms of projects include the Global Fuel Economy with United Nations Environment Programme, Africa Adaptation Program with Government of Japan and regional programme for the sustainable management of the coastal zones of Indian Ocean countries (ReCo Map) among many others. As a result of these partnerships there has been development of sector policies with respect to Land, Biodiversity, Forests, Solid Waste, Coastal Zone Management, tourism and Energy among others.

## A New Concept: The Maurice Ile Durable (MID) Concept

In 2008 the Honorable Prime Minister of Mauritius, Dr. the Hon N Ramgoolam, put forward a vision- The Maurice Ile Durable Vision aiming at transforming Mauritius into a world model of sustainable development. The vision encompasses all the recommendations from the international partnerships referred to in the previous section. However apart from the environmental landscape, the economic and social landscape is considered in this model. The MID vision rests on 5 pillars: Energy, Environment, Education, Employment and Equity (the five 'E's).

The MID is spearheaded by the MID Commission operating under the aegis of the Prime Minister's Office and it operates with other Ministries for implementation of projects. The broad strategic areas being developed under the different pillars are as shown in Table 4.1.

### Table 4.1: Strategic Areas for MID

| Pillar | Strategic Areas |
|---|---|
| Energy | Long Term Energy Strategy, Energy Efficiency Target, Sustainable Public Transport, Renewable Energy Targets, Energy Security, Power Sector Reform |
| Environment | Land Use/Environment, Capacity to increase environmental knowledge, pollution control, water resources, managing wastes, environmental health, improving the environment, food security, biodiversity. |
| Employment/Economy | Develop a Green Economy, Quality of Employment, Sustainable Consumption and Production, Ocean Economy, Human Resources |
| Education | Lifelong learning, Sustainable lifestyles, natural disaster and climate change awareness, access to post secondary education |
| Equity | Support vulnerable groups, Excluded Groups, Institutional and legal framework, Equal Access to infrastructure, arts and culture, Healthy living, individual rights, transparency index |

*Source*: Ministry of Environment and Sustainable Development (2013).

A number of actions under each strategic area has already been initiated.

The concept can be extended to other SIDS through the setup of a science framework providing a multi-lateral forum for international cooperation, information exchange, and dialogue.

## Scientific Partnerships with Africa

Mauritius has over the years built a number of important building blocks for mutually beneficial political and economic relations with Africa.Mauritius has strongly stated its vision and determination to leverage on the opportunities and growth of the African economic space for its own benefit and to enhance its role as a gateway to Africa. Various initiatives have already been spelt out in this direction (Mauritius Africa Club, 2013). In terms of S&T Mauritius is already active in a number of regional initiatives, whether in terms of participation in regional programs, ratification of regional cooperation agreements and offer of scholarships to African students. The Table 4.2 outlines some of the initiatives that Mauritius is already involved in.

**Table 4.2: Participation of Mauritius in some Regional Initiatives**

| Regional Instance | Participation of Mauritius |
| --- | --- |
| NEPAD/SANBio (Southern African Network for Biosciences) | Hosting of the SANBio Bioinformatics Node at the University of Mauritius (UoM) since 2009 |
| Square-kilometre-Array-Project-(SKA) being spearheaded by South Africa in collaboration with Botswana, Ghana, Kenya, Madagascar, Mauritius, Mozambique, Namibia, South Africa and Zambia. This regional team won the bid to house this study since May 2012. | Mauritius is part of the team of 8 countries collaborating in the SKA bid. The SKA telescope will be hosted by South Africa and one of the antennas may be located in Mauritius. |
| COMESA Program for Science, Technology and Innovation | The CBBR of the UoM proclaimed as COMESA Center of Excellence since 2013.Government of Mauritius has nominated one National Research Chair as a member of the Innovation Council of the COMESA. |
| SADC STI Program | Mauritius hosted the first SADC Science Engineering and Technology (SET) Week in 2009, the event was co-chaired by South Africa. |
| ICSU Regional Office for Africa | Mauritius has participated in a number of regional initiatives under the ICSU Regional Office for Africa. These include:<br>1. Contribution for Science Plan on Natural and Human-induced Hazards and Disasters in sub-Saharan Africa, sustainable energy; health and human well-being; natural and human-induced hazards and disasters.<br>2. Holding of international field workshop on the utilization of renewable energy in Africa by UoM/MSIRI in 2007 |

Moreover Mauritius has formulated a number of policy measures to consolidate its Mauritius – Africa Strategy. The integration of S&T in these policy measures could help to build trust and understanding with African counterparts.

## Intensifying Strategic S&T Partnerships with Africa

Over the years Mauritius has successfully geared its economy to adapt to international challenges. The economy, which was based on a single crop (sugar cane) economy for many years, has diversified with the development of new sectors such as manufacturing, tourism, financial services, ICT and the seafood hub. The most important economic sectors in Mauritius are textiles, sugar, tourism and financial services. The GDP is generated at 3 per cent in agriculture/fishing, 34 per cent in industry and 63 per cent in the tertiary sector (business, finances, tourism) (Statistics Mauritius, 2012).Until the 1970's sugar was the primary product for exportation from Mauritius. As sugar prices on the world market continued to fall and international competition increased, economic diversification was needed. Textiles then became the next pillar of the Mauritian economy. Tourism has grown consistently over the past two decades. Information and Communication Technology is regarded as the industry of the future for Mauritius. The island has become a well-known and respected provider of Business Process Outsourcing Services. Financial Services and Offshore Banking is also gaining major importance in the Mauritius economy.

The expertise and know-how of Mauritius in fields such as Agriculture, Textiles and ICT could attract interest on the main land, furthermore addressing regional concerns such as disaster risk and climate change in a collaborative manner would benefit both parties. It is worth mentioning that the formulation of Africa S&T initiatives hold an important place in the forthcoming Science, Technology and Innovation Policy and Strategy (2014-2025) of the MoTESRT.

## Environmental Sustainability; Climate Change and Disaster Risk

The effects of climate change are apparent in Mauritius through rising sea levels (3.8 mm/year), temperature rise of about 10C over the last 50 years and reduced rainfall (MoESD, 2013). Being a Small Island Developing State the susceptibility of Mauritius is exacerbated by its vulnerability of disaster risks among other factors. Being part of Africa Mauritius is also among those countries that are bearing the brunt of climate change despite having the lowest emissions of carbon dioxide (Table 4.3).

**Table 4.3: Emissions of Carbon Dioxide**
**(Billions of metric tons in different parts of the world)**

|  | World | Developing Countries | Developed Countries | Africa | Latin America and Caribbean | Southern Asia | Eastern Asia |
|---|---|---|---|---|---|---|---|
| 1990 | 21.7 | 6.7 | 14.9 | 0.7 | 1 | 1 | 3 |
| 2009 | 30.1 | 16.9 | 13.2 | 1.2 | 1.6 | 2.8 | 8.3 |
| 2010 | 30.7 | 18 | 13.7 | 1.3 | 1.6 | 3 | 9 |

*Source*: The MDG Report 2013, United Nations.

Emissions of greenhouse gases have a global impact, unlike some other forms of pollution. Whether they are emitted in Asia, Africa, Europe, or the Americas, they rapidly disperse evenly across the globe. This is one reason why efforts to address climate change have been through international collaboration and agreement and more so in the region given the increased risk and resilience factor.

International partnership to address climate change has already progressed with the UN Climate Change Conference in Doha, Qatar which forged a consensus on a second commitment under the Kyoto Protocol (2013-2020). Apart from having signed a number of climate-related multilateral environmental agreements, Mauritius has also participated in a number of regional and international projects such as the Climate Change Information Centre (Japan International Cooperation Agency),Capacity Development on Climate Change Measures in Mauritius (Japan International Cooperation Agency) and the Africa Adaptation Programme (AAP) in partnership with UNDP and the Government of Japan (Brief from Ministry of Environment and Sustainable Development, 2013). Mauritius is one of the four countries selected to implement the UNESCO initiative on Climate Change Education for Sustainable Development, the program is aimed to educate young people on the impact of global warming while encouraging changes in attitudes and behavior to promote sustainable development.

At the regional level there is a clear feeling need for more action among African counterparts.This could be by facilitating the sharing of technical know-how and pooling of the necessary laboratory resources and a science framework to provide a multi-lateral forum for regional cooperation, information exchange, and dialogue. Closely linked with climate change there is the issue of Disaster Risk. Developing regions are particularly vulnerable to disasters such as storms, floods, droughts, heavy downpours, cyclones and heat-waves. These are expected to increase with climate change.

Integrated, multidisciplinary research and assessment and enhanced communication between countries can support informed decision-making. Links with Centers of Excellence should be promoted to enhance capacity building activities including, advanced training, visiting fellow programs, training workshops, curriculum development and international conferences.

In such international programs it is worth mentioning that the sharing of data among scientists is a prerequisite. This entails the required expertise and technology to collect data, access data, manage and organize data in a readable form as well as to archive data.

## Information Technologies

Information technologies can play a key role in improving the quality of governance and public administration. The Government of Mauritius is placing much emphasis to modernise its activities and services through e-government efforts, some of the initiatives taken by Government to integrate new technologies in its activities are:

☆ use of educational technology through the Sankoré project in primary schools;

☆ the introduction of the new Identity Card System using smart card technology.

☆ the Human Resource Management Information System project to be implemented by Government will digitalise core human resources; payroll; performance management; learning management and employee self-service; and

☆ online tax collection.

☆ online driver's licence applications

☆ computerised land administration system for improved urban planning, infrastructure development, environmental management and production of national statistical data through the LAVIMS (Land Administration Valuation and Information Management System) project.

☆ The experience of Mauritius in this field could be shared with countries from Mainland Africa through capacity building and joint programs.

Moreover in the field of ICT the African Union is gearing up to discuss a brand new cyber security initiative for the continent (African Union Commission, 2012). African countries have recently suffered an increase in phishing scams, malware, advance-fee scams and mobile-money related fraud, according to industry insiders. In this line it is noteworthy that Mauritius has developed a National Cyber Security Strategy (2014-2019) (Ministry of Information and Communication Technologies) to effectively protect information systems and networks. Hence Cyber Security is a field that is of common interest to Mauritius and Mainland Africa and could be further developed through science cooperation programs.

## 4. Conclusions

It is clear that S&T are tools which can be used to enhance relationships between Mauritius and other countries. However it is also clear that systems should be established in which Mauritius and its counterparts can mutually benefit so that the right synergy will be built between S&T and Diplomacy for resolving global issues. Along this line the appropriate 'Human Resource Capacity' should be developed to sustain these efforts by having a particular blend of scientific and diplomatic expertise. The possibility of integrating science expertise in foreign postings for informed diplomatic actions should be explored. These mechanisms will definitely contribute towards the emergence of Mauritius as a role model and key player among SIDS and in Africa.

## 5. Acknowledgements

The authors wish to express their appreciation to the following institutions:

☆ The Ministry of Environment and Sustainable Development, Government of Mauritius for having kindly shared information on international initiatives related to Sustainable Development.

&#9734; The Mauritius Research Council for funding the participation of the Representative from MRC in the workshop.

&#9734; The Centre for Science and Technology of the Non-Aligned and Other Developing Countries (NAM S&T Centre) for partial support to participate in the workshop.

## 6. Abbreviations

AAP: Africa Adaptation Program

BPoA: Barbados Programme of Action

CBBR: Centre for Biomedical and Biomaterials Research

COMESA: Common Market for Eastern and Southern Africa

ICSU: International Council For Science

ICT: Information and Communication Technologies

IOC: Indian Ocean Commission

MID: Maurice Ile Durable

MoESD: Ministry of Environment and Sustainable Development

MoFARIT: Ministry of Foreign Affairs, Regional Integration and International Trade

MoTESRT: Ministry of Tertiary Education, Science, Research and Technology

MSIRI: Mauritius Sugar Industry Research Institute

MRC: Mauritius Research Council

NEPAD: New Partnerships for Africa's Development

S&T: Science and Technology

SADC: Southern Africa Development Community.

SANBio: Southern African Network for Biosciences

SIDS: Small Island Developing State

R&D: Research and Development

SKA: Square-kilometre-Array

UNDP: United Nations Development Program

UNESCO: United Nations Educational, Scientific and Cultural Organisation

UoM: University of Mauritius

## REFERENCES

1. African Union Commission (2012). Draft African Union Convention on the establishment of a legal framework conducive to Cyber Security in Africa. African Union.

2. Flink, T. and Schreiterer, U. (2010). Science Diplomacy at the intersection of S&T policies and foreign affairs: toward a typology of national approaches.

3.  Government Information Service (2012). Government Information Service Newsletter, June 2012. Government of Mauritius.

4.  Government Of Mauritius (2012). Government Programme (2012-2015)-Moving the Nation Forward. Government of Mauritius.

5.  Mauritius Africa Club (2013). Strengthening Economic Growth of Mauritius through a coherent, deepened and effective Mauritius. Mauritius Africa Strategy, An advocacy paper.

6.  Ministry of Environment and Sustainable Development (2013). Maurice Ile Durable: Policy, Strategy and Action Plan. Republic of Mauritius

7.  Ministry of Finance and Economic Development (2014). Budget Speech 2014-Building a better Mauritius.

8.  Ministry of Information and Communication Technology (2013). National Cybersecurity Strategy (2014-2019).

9.  Nye J (2004) Soft Power: The means to Success in World Politics.Public Affairs: New York

10. Sunami, A., Hamachi,T. and Kitaba, S. (2013). The Rise of Science and Technology Diplomacy in Japan. A Quarterly Publication from the AAAS Center for Science Diplomacy.

11. Prime Minister's Office (2013). The Ocean Economy-A Road Map for Mauritius, Prime Minister's Office, Government of Mauritius

12. Statistics Mauritius (2012). Mauritius in Figures. Ministry of Finance and Economic Development, Republic of Mauritius.

13. The Royal Society (2010). New Frontiers in Science Diplomacy. Navigating the changing balance of power.

14. United Nations (2013).The Millennium Development Goals Report 2013

15. Whitesides, J. (2011). Better Diplomacy, Better Science. China Economic Review, Oct 2011.

16. Yakushui, T. (2009). The potential of Science and Technology Diplomacy. Asia-Pacific Review. 16(1):1-7.

*Chapter 5*

# Status of Science and Technology Diplomacy and Need for Capacity Building in Nepal

*Chiranjivi Regmi*

*Nepal Academy of Science and Technology,*
*Khumaltar, Lalitpur, Nepal*
*e-mail: info@nast.org.np; planning@nast.org.np*

## ABSTRACT

Science and Technology diplomacy is the use of scientific collaborations among nations to address common problems and to build constructive international partnerships. New emergent forms of international diplomacy are developing to deal with a number of emerging issues in which science and technology play a central role. As science and technology are increasing at the center of global issues, diplomats are less capable of effectively completing their works without heavily relying on scientists and engineers for clarification and insight.

Exposure, peace and diplomatic relationship of a country with outside world has always been playing a vital role to share and introduce technologies.

The partnership between United Nations Organization and Nepal started as early as 1950 when Mr. Tony Hagen- Swiss National was appointed by the United Nations Technical Assistance Administration (TAA) to carry out a survey of Nepal's Mineral Resources. Mr. Hagen who was the first outsider to visit the whole territory of the country during 1950-1958 prepared "Geological Report on Nepal". He also made observation reports on general nature of the country, the ethnology and life of the people and the economy and the potentialities for development. These reports have been regarded as very important documents for the exposure of the country and mobilise support for science and technology cooperation.

The fundamental objective of the Nepal foreign policy is to enhance the dignity of Nepal in the international arena by maintaining the sovereignty, integrity and independence of the country.

Nepal has established Diplomatic Relations with 132 countries in the world. It is an active member of the Non-Aligned Movement, the United Nations, its Specialised Agencies as well as other International Organizations, and a founding member and active player in the South Asian Association for Regional Cooperation (SAARC).

In 1976 the National Council for Science and Technology (NCST) was created to formulate S&T policy and coordinate S&T activities in the country.

In 1982, the (Royal) Nepal Academy of Science and Technology was established as an autonomous apex body of science and technology for the all-round development of the country and in fact replaced NCST with broader objectives and legislation. After the establishment of Parliamentary Democratic system in 1990 the Parliament promulgated RONAST Act in 1992 and amended it as NAST Act in 2007 after the establishment of Republican system in Nepal.

In 1996, the government established the Ministry of Science and Technology now renamed as Ministry of Science, Technology and Environment (MoEST). The MoEST is the link ministry of NAST and they are working together in different aspects of science and technology development in the country. Institute of Foreign Affairs (IFA) was founded as an integral part of the Ministry of Foreign Affairs in 1993. IFA works with its partner organizations, governmental as well as nongovernmental organizations, at home and abroad, to realize its objectives relating to foreign policy issues and its interactions with political, economic, social and cultural aspects.

Training scientists, diplomats and other stake holders in science and technology diplomacy, policy formulation and regulatory matters is the urgent need of today.

It is felt that trainings are required in areas of science and technology diplomacy such as; Code of conduct for technology transfer, Acquisition and absorption of new technologies, Technology partnership, Intellectual Property Rights and protection of traditional knowledge and Biotechnology and information and communication technologies, Science, technology and innovation policy, Science and Technology capacity development.

*Keywords: Science, Technology, Diplomacy, Innovation.*

# 1. Introduction

At present, different countries of the world grouped as "Developed, Developing and Least Developed countries on the basis of their economic development. Differences between them are basically due to their differing mastery and the utilization of the latest science and Technology. Nepal belongs to the group of Least Developed Countries (LDC).

Exposure, peace and diplomatic relationship of a country with outside world has been always playing a vital role to share and introduce more advanced technologies than the indigenous ones that are useful for the country. Modern Nepal was founded in 1743 by the king Prithvi Narayan Shah following the unification of different petty states. Nepal engaged for about a century in conflicts with Tibet and British Colonial Empire in India. Another full century of complete isolation followed with the establishment of Rana autocracy in 1846. For these two centuries, Nepal had little contact with outside world and virtually remained isolated from the rest of the world 9. After the visit of Prime Minister Mr. Chandra Shumsher JB

Rana to Great Britain, three visible landmarks of modern technology ; Bir Hospital, Hydroelectric power plant at Pharping near Kathmandu and a rope line were built during 1890-1920 (Bajracharya, 2001). It is one of the first examples of science and technology diplomacy.

Nepal remained almost totally unaffected by the development of science and technology in the world till the middle of the 20th century. Institutions of modern science and technology had begun to be established only after the end of the Rana Regime and the advent of democracy in 1951.

The partnership between United Nations and Nepal was started as early as 1950 when Mr. Tony Hagen- Swiss National was appointed by the United Nations Technical Assistance Administration (TAA) to carry out a survey of Nepal's Mineral Resources. Mr. Hagen who was the first outsider to whole territory of the country during 1950-1958 prepared "Geological Report on Nepal". He also made observation reports on general nature of the country, the ethnology and life of the people and the economy and the potentialities for development. These reports have been regarded as very important documents for the exposure of the country and mobilise support for science and technology cooperation (Tony Hagen, 1959 and 1960).

The United Nations organization has been working with the government and people of Nepal since the early 1950s to promote and preserve the basic rights outlined in the charter for peace, security and development. Number of UN agencies such as UNEP, UNESCO, WHO, FAO, WMO, WIPO, and ITU are dealing with science and technology issues are working in Nepal in regular basis. In addition Treaties such as UN Framework Convention on Climate Change (UNFCCC), The Convention on Bio logical Diversity (CBD) and the Convention to Combat Desertification (CCD) World Trade Organisation (WTO) are providing opportunities for the involvement of Government and Non-Government Organisations,Academies and research institutions to share with these organisations related to science and technology (Regmi *et al.*, 2010)

## 2. Guiding Principles of Foreign Policy

Nepal has established Diplomatic Relations with 132 countries in the world. It is an active member of the Non-Aligned Movement, the United Nations and its Specialised Agencies as well as other International Organizations, and a founding member of an active player in the South Asian Association for Regional Cooperation (SAARC).

The fundamental objective of the foreign policy is to enhance the dignity of Nepal in the international arena by maintaining the sovereignty, integrity and independence of the country. The foreign policy of Nepal is guided by the abiding faith in the United Nations and policy of nonalignment.

The basic principles guiding the foreign policy of the country include: Mutual respect for each other's territorial integrity and sovereignty; Non-interference in each other's internal affairs respect for mutual equality Non-aggression and the peaceful settlement of disputes Cooperation for mutual benefit.The foreign policy of Nepal is also guided by the international law and other universally recognized norms governing international relations (United Nations Conference on Trade and Development, 2003).

The diplomatic relations of the country is not limited to the national sovereignty, territorial integrity and nationality. In the present day, it has buckled-up with the economic, security, social, environment, and other interests of the countries. Therefore, the issues such as economic benefits and security come at the front while talking about diplomatic relations of any countries. Now, uninterrupted socio-economic progress of the people is at the core of the foreign relations through economic diplomacy. It has been realized that the conduct of foreign relations is at its best when a country is economically strong with favourable external environment for the pursuit of collective prosperity and the individual welfare of all the Nepalese people.

## 3. Science Diplomacy

New forms of international diplomacy are developing to deal with a number of emerging issues in which science and technology play a central role.

Science diplomacy is the use of scientific collaborations among nations to address common problems and to build constructive international partnerships. Science diplomacy has become an umbrella to describe a number of formal or informal technical, research-based, academic or engineering exchanges (United Nations Conference on Trade and Development, 2003). The science diplomacy might have following meanings:

- ☆ **Science in diplomacy**: Science can provide advice to inform and support foreign policy objectives.
- ☆ **Diplomacy for science**: Diplomacy can facilitate international scientific cooperation.
- ☆ **Science for diplomacy:** Scientific cooperation can improve international relations.

The UN Conference on Trade and Development (UNCTAD) serves as a focal point with in UN secretariat for all matters related to foreign direct investment, transitional cooperation, enterprise development, and science and technology for development. UNCTAD was given responsibility to "develop special programme to contribute for training scientists diplomats and journalists in science and technology diplomacy, policy formulation and regulatory matters to assist developing countries, in particular least developed countries, in international negotiations, and international norms and standard setting". This was the outcome of the resolution (2001/31) that was adopted by the UN Economic and Social Council (ECOSOC) in July 2001, following recommendation of the UN Commission on S&T for Development (UNCSTD) and consultation with the Secretary General of UN Conference on Trade and Development to develop special programme on Science Technology Diplomacy.

There are two key feature of the growth of scientific and technological knowledge that are central to international negotiations. Firstly, scientific knowledge is becoming increasingly specialized and therefore demands greater expert input into international negotiations. Secondly, the application of science and technology requires developing the ability to integrate the divergent disciplines that are needed to solve specific problems. International diplomacy demands that government negotiators deal with both specialization and integration.

## 4. Institutions Related to S&T Diplomacy

After the advent of Democracy in 1951 several departments, related to science and technology, were established. They were oriented towards the implementation of science and technology development programmes. However there was lack of organised science and Technology activities in the country. In 1976 the National Council for Science and Technology (NCST) was created to formulate S&T policy and coordinate S&T activities in the country. It worked as a focal point for the development and application of science and technology in the country. However it could not achieve its goals as expected.

### Nepal Academy of Science and Technology (NAST)

In 1982, the (Royal) Nepal Academy of Science and Technology was established as an autonomous apex body of science and technology for the all-round development of the country and in fact replaced NCST with broader objectives and legislation. After the establishment of Parliamentary Democratic system in 1990 the Parliament promulgated RONAST Act in 1992 and amended it as NAST Act in 2007 after the establishment of Republican system in Nepal. The objectives of the Academy are; Advancement of science and technology for all – development of the nation, preservation and further modernization of indigenous technologies, promotion of research in science technology, identification and facilitation of appropriate technology transfer. The main activities are ; Advise the Government of Nepal (GoN) on science and technology development programmes on the formulation of technology transfer policy and establishment of new institutions or laboratories for S&T related research and development, Establish and strengthen relationship with regional and international institutions in order to promote mutual cooperation, mobilise internal and external both technical and financial resources, for science and technology development and implement and S&T programmes in collaboration with national and international organisations. With regards to the formulation of the National Science and Technology Policy, (RO) NAST prepared a draft and presented in the First Conference on Science and Technology in 1988 and got endorsed by the science and technology community of the country. The government had endorsed it and published in Nepal Gazette. In fact it was the first National Policy Document on Science and Technology in the country.

NAST has been working as the focal point of several International organisations such as Third World Academy of Sciences (TWAS), International Council of Scientific Unions (ICSU), International Development Research Center (IDRC), International Foundation of Sciences (IFS), The Centre for Science and Technology of Non- Aligned and Other Developing Countries (NAM S&T Centre), Association of Science Academies and Societies in Asia (ASSA) and Science and Technology Policy Asian Network (STEPAN). NAST has developed bilateral Memorandum of

Agreement with more than 15 research organisations and academies of different countries. Indian National Science Academy (INSA), Council of Scientific and Industrial Research (CSIR), India, Chinese Academy of Sciences (CAS), China, NIMS, Japan, Academia Sinica, Taiwan, National Research Council (CNR), Italy, Royal Botanical Garden, Edinburgh, UK, National Science and Technology Development Agency(NSTDA), Thailand are some the organisations with those that NAST has active collaboration. The GoN has given responsibility of developing strategic plans for the development of capability of the country in the field of science and technology.

In 1996, the government established the Ministry of Science and Technology now renamed as Ministry of Science, Technology and Environment (MoEST). The ministry developed science and Technology Policy in 1997 in collaboration with NAST. The MoEST is the linkage ministry of NAST and they are working together in different aspects of science and technology development in the country. The MoEST is always seeking advice and cooperation from NAST keeping in view with the asset of scientific capability that exists at NAST in terms of human resources and laboratory facilities.

## Institute of Foreign Affairs (IFA)

Institute of Foreign Affairs (IFA) was founded as an integral part of the Ministry of Foreign Affairs in 1993; it got reconstituted as a semi-autonomous body in the year 2012. Honourable Minister of Foreign Affairs heads the board of directors of the institute and the Foreign Secretary represents the ministry as a member to the board.

The objectives of IFA are:

☆ Provide suggestions and recommendations to Government of Nepal (GoN) on short and long term policy formulation'

☆ Prepare concept papers including strategic analysis on foreign affairs and submit to GoN'

☆ Organize seminars workshops, meetings, and conferences to discuss and deliberate on foreign policy issues and come up with recommendations

☆ Undertake study and research programmes on more pressing foreign policy issues and objectives'

☆ Provide training to officials of the Ministry of Foreign Affairs and other Ministries on foreign policy issues and objectives'

☆ Update and systematically compile all historical documents and information on foreign policy issues and publish them as and when necessary'

☆ Establish linkage with foreign governments, INGOs, NGOs and eminent personalities to achieve common objectives

The institute is expected to engage the services of eminent scholars, foreign policy experts, diplomats, ambassadors, and other experts on international relations in the construction of long term foreign policy goals and objectives.

The Ministry of Foreign Affairs and Diplomatic Missions are engaged in performing the Political Diplomacy, Cultural Diplomacy, Military Diplomacy and more recently Economic Diplomacy. However, there has never been spelled out about Science Diplomacy in Nepal so far (Gautam, 2013)

To address the Economic Diplomacy some activities have been initiated. The new areas for Foreign Direct Investment (FDI) are opened up thorough the FDI Act 1992. More recently the GoN has taken the initiative to establish "Special Economic Zone" with a view to attracting FDI. In total 116 joint ventures comprising of manufacture, services, tourism, construction, agriculture, mineral and energy were permitted to operate FDI during FY 2005/06. However the country has not become successful to attract FDI as planned. The nation possesses 2.3 percent of world's hydro resources, magnificent biodiversity resources and huge potentiality for tourism development. Preliminary geological surveys reveal that several mineral resources are available in the country. It is possible to attract FDI in such areas as hydropower, tourism, biodiversity and education. Diplomatic missions need to be mobilized for the promotion of FDI.

# 5. Role of Science and Technology Diplomacy

There are some discussions among the Nepalese diplomats that Nepal should stop the begging- bowl mentality. To attract international support, progress made by the country with the development of Science and technology capability has to be recognized.

## Science and Technology in Diplomacy

The country has to develop a long term reconstruction and development plans. At present, Nepal has already developed critical mass of scientists and technologists capable of developing plans for infrastructure development. There is a need for political commitments and vision. Based on such long-term plans the country can mobilize international support.

There are two key features of the growth of scientific and technological knowledge that are central to international negotiations. Firstly, scientific knowledge is becoming increasingly specialized and therefore demands increase expert input into international negotiations. Secondly, the application of science and technology is to develop the ability to integrate the divergent disciplines that are needed to solve specific problems. International diplomacy presently demands that government negotiators deal with both specialization and integration.

As science and technology are increasing at the center of global issues, diplomats are less capable of effectively completing their work without heavily relying on scientists and engineers for clarification and insight. The effectiveness and international civil servants increasingly depend upon the extent to which they can mobilize scientific and technical expertise in their work. Although a number of experts are available, they are not designed to receive systemic science advice as the basis for effective performance.

## Science and Technology Diplomacy and MDGs

With regards meeting the Millennium Development Goals (MDG) Nepal has made commendable progress in reducing maternal and child mortality, increasing girls' education, controlling diseases, improving access to water supply and sanitation, immunization, community forestry etc. Nepal consist of eradication of poverty and hunger from 38 percent to 21 percent by 2015, attainment of 100 percent universal primary education, 100 percent promotion of gender equality and women empowerment, combating HIV/AIDS malaria and other diseases ensuring, environmental sustainability and developing a global partnership for development. In this way application of Science and Technology significantly contributes towards meeting MDGs (Anonymous, 1999)

## Science and Technology in Bridging the Neighboring Countries

Nepal's new foreign policy agenda has appeared recently to address the economic diplomacy. It is about the defining new role of Nepal to develop economic bridge between India and China. Nepal's landlocked geographic position and its size until few years ago was perceived as a major handicap of country's overall development but speedy development of these two neighbors has indicated to shift Nepal's Foreign Policy Agenda from traditional towards becoming the vibrant bridge. Some scholars mention it shifting from geo-politics to geo-economics. Preliminary activities have been initiated towards this, but more has to be done and it could be through Science and Technology Diplomacy.

## Climate Change and Science and Technology Diplomacy

The executive director of IFA, has reported that the geographic location of Nepal is in the forefronts of acute vulnerability to the impacts of climate change crisis with threats posed by the melting glaciers of the Himalayas. Many of the region's major river systems provide it with strategic climate change adaptation opportunities to monitor and regulate river flows (Adhikari, 2012).

Nepal can develop and implement long term plans or joint projects (India, Bangladesh, Nepal and Bhutan) for climate change and security in South Asia through developing skills in Science and Technology Diplomacy.

Not only that the Mt Everest region is regarded as the Third Pole and many scientists are interested to carryout research work on climate change in this region. There are tremendous opportunities for exploring scientific cooperation and side by side economic betterment of the country. Similarly Climate diplomacy will provide opportunities for Carbon trading.

## Science and Technology in Global Trading System

Developing countries have raised concerns over the impact of International Agreements, on their abilities to strengthen their productive capacity through trade related industrial and investment measures. More specifically these countries are concerned about the degree to which they can formulate policies that enhance technological development without contravening WTO rules. Similarly the Agreement on Trade –related Aspects of Intellectual Property Rights (TRIPS) also needs good capability of a nation to protect IPR. There is a clear relationship between technological and regulatory capacity including standard setting. Standards are major determinants of trade flows and policies and as such may have positive and negative effects. However, Nepal, like many developing countries, has remained "Standard –taker" rather than "Standard-maker" so far.

## 6. Training Needs

In an interview the Director of IFA, Dr. Rishi Adhikary stated "IFA will continue to conduct seminars and interactions on relevant topics and publish and share with all the concerned stakeholders for raising awareness. It has plans to extend relationships with similar organizations in South Asian Association on Regional Cooperation (SAARC) member and observer countries. IFA also plans to have more effective coordination with MoFA and other relevant organizations for better delivery". It shows the need of Training on different aspects of Science and Technology

Training scientists, diplomats and other stake holders in science and technology diplomacy, policy formulation and regulatory matters is the urgent need of today.

It is felt that trainings are required in following areas of science and technology diplomacy;

- ☆ Code of conduct for technology transfer
- ☆ Acquisition and absorption of new technologies
- ☆ Technology partnership
- ☆ Intellectual Property Rights and protection of traditional knowledge
- ☆ Biotechnology and information and communication technologies.
- ☆ Science, technology and innovation policy
- ☆ Science and Technology capacity development
- ☆ Climate Change and Diplomacy
- ☆ Carbon Trading

## 7. Acknowledgment

I extends my sincere thanks to Prof. Surendra Raj Kafle, Vice Chancellor of NAST and to Director, NAM S&T Center Prof. Dr. Arun Kulshreshtha for giving me the opportunity to participate in this very special and important international workshop. Similarly I would like to sincerely thank to my colleagues at NAST specially Mr. Niranjan Acharya and colleagues at NAM S&T Center specially Mr. M. Bandyopadhyay for their cooperation and valuable suggestions.

## REFERENCES

1. Adhikari, R.R., 2012. Climate Change and South Asia. Climate Change as a security risk in South Asia. IFA p.98-100.

2. Anonymous, 1999. Nepal; Common country Assessment. The United Nations System, Kathmandu, Nepal.

3. Bajracharya, D.,2001. Science and Technology in Nepal RONAST, Kathmandu, Nepal, pp. 214.

4. Gautam, K.C., 2013. Enhancing Effective Participation of Nepal in International System. In Foreign Policy of Nepal. IFA p15-55.

5. Institute of Foreign Affairs, 2011. Speeches of Heads of the Nepalese delegations to the Non-Aligned Movement (1961-2009).

6. Institute of Foreign Affairs, 2013. Institutionalization of Nepal's Foreign Policy.

7. Regmi,C., I.P.Khanal, S.K.Desar, V.Singh, D.K.Shah,and G.K.Pokhrel, 2010. Status of Science and Technology in Nepal NAST, Kathmandu Nepal.pp.175.

8. Tony Hagen, 1959 reprinted by UNDP in 1997,Observations on Certain Aspects of Economic and Social Development Problems in Nepal.

9. Tony Hagen, 1960 reprinted by UNDP in 1997,A brief Survey of the Geology of Nepal.

10. United Nations Conference on Trade and Development, 2003. Science and Technology Diplomacy- concepts and Elements on a work Programme, United Nations, New York and Geneva.

## Chapter 6

# Science and Technology Diplomacy: Impacts, Achievements, Opportunities and Challenges

*Clifford Mupeyiwa*

**Principal Science and Technology Officer,
Ministry of Higher and Tertiary Education
Science and Technology Development,
Zimbabwe
e-mail: jenahxp@gmail.com**

## ABSTRACT

The Department of Science and Technology Development (DSTD), within the Ministry of Higher and Tertiary Education Science and Technology operates within the framework of the Science and Technology (S&T) Policy with a mandate to promote, facilitate and coordinate the strategic application of S&T into the mainstream of the economic activities in Zimbabwe. In order to achieve the above goal the DSTD works with all S&T related sectors of the economy chief among them; Agriculture, Energy, Research and Development (R&D) Institutes, Mining, Information and Communication Technology (ICT), Intellectual Property (IP) Organisations, Small to Medium Enterprises, Education, Health, Environment and Finance.

This paper seeks to present the impact, challenges and opportunities presented by use of S&T Diplomacy to break political impasses by ensuring collaboration of all global nations through use of scientific solutions to local regional and global challenges using available material, technologies and capital and S&T human resources. The term "science and technology diplomacy" is used to mean the provision of science and technology advice to multilateral negotiations and the implementation of the results of such negotiations at the national level. It therefore covers activities at the both international level and national level pursuant to international commitment. Advances in science and technology have become key drivers in international relations, and knowledge of trends in key fields is an essential prerequisite to effective international negotiations. Knowledge of trends in science and technology is also a key element for the successful national implementation of international agreements. There are two key features of the growth of scientific and technological knowledge that are central to international negotiations. Firstly, scientific knowledge is becoming increasingly specialized and therefore demands greater expert input into international negotiations. Secondly, the application of science and technology to development requires the ability to integrate the divergent disciplines that are needed to solve specific problems. International diplomacy now demands that government negotiators deal with both specialization and integration.

The main focus of the presentation would be on the continual lobbying and buy-in for realignment of S&T, as well as Foreign Policies of developing nations, so as to be able to sustainably leap frog their economies through S&T Diplomacy and be able to achieve quick wins through initiatives in IPR, Commercialisation, exchange of information, capacity building and for technology sourcing and building of S&T partnerships. Areas of leapfrogging developing countries would be in emerging technologies of Nanotechnology, Biotechnology, Indigenous Knowledge Systems (IKS), Robotics, Mechatronics and Information Communication Technologies (ICTs). Finally a review of the current status of S&T diplomacy in various developing countries through identification of training needs and recommendation for suitable mechanisms of cooperation through sharing the capabilities and experiences of the developing countries on S&T diplomacy would be of necessity.

*Keywords*: Science, Diplomacy, Science diplomacy, Impacts, Opportunities and challenges.

# 1. Introduction

Zimbabwe is located in southern Africa. According to Sibanda (2011), Zimbabwe has a land area of 390,759 sq. km (150,873 sq. mi). From north to south its greatest distance is 760 km (470 mi), and from east to west it is 820 km (510 mi). The country's east is mountainous with Mount Nyangani as the highest point at 2,592 m. About 20 per cent of the country consists of the low veld under 900m. Victoria Falls, one of the world's biggest and most spectacular waterfalls, is located in the country's northwest as part of the Zambezi river. The country has a tropical climate with a rainy season usually from late October to March. The climate is moderated by the altitude. Zimbabwe is faced with recurring droughts; and severe storms are rare.

The country borders Mozambique to the east and Botswana to the west. South Africa is located to the south, and the Limpopo River forms the boundary between the two countries. In the north the border is formed by the Zambezi River, beyond which is Zambia. The map in Figure 6.1 shows the location of Zimbabwe in Africa.

Zimbabwe has two major language dialects, Shona and Ndebele. Shona is the major dialect spoken by about three quarters of the Zimbabwean population which covers areas in the south, east and west of the country. The Ndebele occupy the western parts of Zimbabwe {7}.

**Figure 6.1: Location of Zimbabwe in Africa.**

## 2. Evolution of Science in Zimbabwe

### 2.1 Colonial Period

During the Rhodesian era, it was a time of low level technological transformation. Innovations and technologies wear geared to the immediate requirements of agriculture and industry. The history of SETI (science, engineering, technology and innovations) in Zimbabwe dates back to 1967 when the then Prime Minister of Rhodesia established Scientific Liaison Office in the Office of the Prime Minister. The office was tasked with the duty of advising the Prime Minister on scientific matters. The Rhodesian government established the Rhodesian Iron and Steel Commission (RISC), Cotton Industry and Research Board and Industrial Development Commission with the purpose of stimulating ideas for management of the industrialised economy but there was no sign of technology development. During the UDI (Unilateral declaration of Independence) era there was no explicit S&T(Science and Technology) policy document. {5}

### 2.2 Post-colonial Period

It took close to 22years for Zimbabwe to produce the 1st S&T policy document. Challenges retarding the process included; lack of a strong policy community making it impossible to make well defined plans, the formulation process was continuously being shifted to different ministries thus delaying the process. The policy making process was also a bit complex due to lack institutional coherence and synergism. Other factors delaying the process included lack of funds, lack of consultation, participative and inclusive approach. {1}.The first S&T policy was launched by His Excellence the President Comrade Robert Gabriel Mugabe in 2002, and subsequently followed in 2004 with the establishment of a Department of Science and Technology (DST) under the Office of President and Cabinet. In the same year 2004, due to the intensive advocacy for funding of inventions ready for commercialisation by ZAI (Zimbabwe Association of Inventors) which had been born a year earlier than the policy in 2001, an Innovation and Commercialisation Fund (ICF) was established. The DST was promoted to a fully-fledged Ministry of Science and Technology Development (MSTD) in 2005 as an implementation to S&T Diplomacy influenced at regional level by the Southern African Development Community (SADC). The S&T policy was reviewed in 2012 into a Science, Technology and Innovation (STI) Policy. The reviewed STI Policy of 2012 was in line with global S&T trends, being crafted through S&T local, regional and global partners at very highest levels of SADC and UNESCO, as well as competent and highly skilled Zimbabwean S&T human capital based in other countries. In summary the STI policy focuses on 6 main goals of: S&T capacity development; learning and utilizing emergent technologies in accelerating development; search for scientific solutions to global environmental challenges; mobilize resources ad popularise S&T; and foster international collaboration in STI.{11}

## 3. Defining Science Diplomacy

The growing complexity of science and innovation systems and the interface with society have been accompanied by a more complex policy environment. This results in a need for better coordination and coherence at national level. One the most crucial factors is the increasingly global nature of the issues with which national policy-makers are confronted. In a whole series of areas such as the environment, telecommunications, health, energy, education and intellectual property, it no longer makes much sense to construe problems in purely sectorial and national terms. In a world that is becoming daily more interdependent, policy-making is inevitably assuming an increasingly transversal and global dimension. In this context, science, technology and innovation (STI) policy systems have emerged as interconnections between knowledge, values, national and international socio-economic, environmental, technological and organizational components {11}

Science diplomacy is the use of scientific interactions among nations to address the common problems facing humanity and to build constructive, knowledge based international partnerships.(Dr Nina Federoff, Science and Technology Adviser to US Secretary of State){8}or Generally speaking, science diplomacy is the use of science, its methods, and its philosophies in diplomacy as an avenue for establishing new connections and strengthening existing ones. Science is yet another field that can broaden horizons and diversify the international dialog, handily lending itself to problem solving, logical discourse, and the ongoing pursuit of understanding that diplomacy currently espouses.

The British Royal Society and the American Association for the Advancement of Science (AAAS), describe three major facets of science diplomacy in a 2010 journal focused on providing a concise definition for the term science diplomacy. Science in diplomacy, science for diplomacy, and diplomacy for science are the three pillars that provide a basis for science diplomacy. Science in diplomacy entails science informing and advising foreign policy, ultimately providing a more unimpeachable body of support for any given objective. Science for diplomacy is the notion that science can be used as a diplomatic tool, through the notion of soft power, to shape international dialogs and to create more channels of communication between communities. Finally, diplomacy for science consists of efforts to involve international actors in the pursuit of science. {6}

Perhaps most importantly is the maintenance of the philosophies in each word of the term. Science attempts to unravel the mysteries of the universe through reasoned approach, rigorous testing, and communal review and understanding. Diplomacy seeks to bridge the gaps between the world's communities, employing the pursuit of tolerance and understanding with the ultimate goal of resolving common differences. Science diplomacy primarily seeks to bring these two concepts together so that each individual aspect of their doctrines can enhance the other. {9}

## 4. Impact

Scientific values of rationality, transparency and universality are the same the world over. They can help to underpin good governance and build trust between nations. Science provides a non-ideological environment for the participation and free exchange of ideas between people, regardless of cultural, national or religious backgrounds.

☆ Science is a source of what Joseph Nye, the former dean of the Kennedy School of Government at Harvard University, terms 'soft power' (Nye2004). The scientific community often works beyond national boundaries on problems of common interest, so is well placed to support emerging forms of diplomacy that require non-traditional alliances of nations, sectors and non-governmental organisations. If aligned with wider foreign policy goals, these channels of scientific exchange can contribute to coalition building and conflict resolution.

☆ Science diplomacy seeks to strengthen the symbiosis between the interests and motivations of the scientific and foreign policy communities. For the former, international cooperation is often driven by a desire to access the best people, research facilities or new sources of funding. For the latter, science offers potentially useful networks and channels of communication that can be used to support wider policy goals. {8}

Human-induced global problems that confront us cannot be solved by any one individual, group, agency or nation. It will take a large collective effort to change the course that we are on; nothing less will suffice. Our planet is facing several mammoth challenges: to its atmosphere, to its resources, to its inhabitants. Wicked problems such as climate change, over-population, disease, and food, water and energy security require concerted efforts and worldwide collaboration to find and implement effective, ethical and sustainable solutions. Given the long-established global trade of scientific information and results, many important international links are already in place at a scientific level. These links can lead to coalition-building, trust and cooperation on sensitive scientific issues which, when supported at a political level, can provide a 'soft politics' route to other policy dialogues. That is, if nations are already working together on global science issues, they may be more likely to be open to collaboration on other global issues such as trade and security. {10}

☆ One other impact of science diplomacy is **hope.** Linkages and collaborations within the science sector have given great hope to communities (especially marginalised ones) of the world that the world can be a better place. Science diplomacy has led to development of programmes which takes into cognisance of both developed and developing nations.

## 5. Achievements (Science Diplomacy in Action)

☆ Pugwash Conference on Science and World Affairs is one success story on science diplomacy. It brings together influential policy officials, public figures and scientists to seek ways to eliminate nuclear weapons and reduce threats of war.{3}

☆ SAFARI 2000), which from 1998 to 2003 brought together two hundred scientists across sixteen countries. It was a multinational environmental and remote-sensing field campaign that observed a broad range of phenomena related to land-atmosphere interactions and biogeochemical functioning across southern Africa. Its objective was to better understand how aerosol and trace gas emissions affect local and regional climate and ecosystems. The initiative traced atmospheric emissions from sources to deposition and involved coordinated satellite, aircraft, and ground-based observations during intensive field campaigns and long-term monitoring at core ground sites across the Southern African Development Community (SADC) region. Such regional networks create necessary and enabling conditions for favourable science diplomacy outcomes

☆ The Fulbright program has been in existence in Zimbabwe since 1982 and approximately 80 Zimbabweans have received Fulbright grants to pursue higher degrees or carry out research in the U.S.(Harare, May 15, 2014: The U.S. Embassy announced this week that 30 Zimbabweans have been invited to participate in the first ever Young African Leadership Initiative (YALI) Washington Fellowship. These outstanding young Zimbabwean leaders will travel to the U.S. this June with over 470 other young African leaders for a six week program at one of 20 prestigious U.S. universities followed by a conference in Washington, D.C. hosted by President Obama.) This is despite of the sanctions the country is under. KOICA/JICA also have been existence for a while helping Zimbabwe in capacity building across all sectors especially science{12}

☆ The development of a nanotechnology centre in Zimbabwe in collaboration with Buffalo University (Programme already launched and ground work has already been started)

☆ Technology hub with the support of US Embassy in Harare/Hivos/Indigo trust. The hub is called the Hypercube Hub (the hub natures, develops and identifies best minds in ICTs and supports their developments with resources ranging from funding, exchange opportunities and enabling environments for them to continue sprouting)

☆ The construction of the International Space Station (ISS) began in earnest and dramatic fashion in 1998 when the U.S.-built module Unity was mated with a Russian module using the Canadian-built robotic arm on the space shuttle. After fourteen years of multiple redesigns, cutbacks, and intricate intergovernmental negotiations, the dream of a permanent, peaceful, and collaborative occupation of near Earth orbit had begun. What is perhaps the most complex and technically ambitious large-scale engineering project ever undertaken by a group of nations; the building of a scientific laboratory in the harsh environment of lower Earth orbit—is as much a foreign policy and human achievement as it is a technical one.

## 6. Opprotunities

☆ Development of new scientific partnerships with other global partners *e.g.* the Middle east/Islamic world

☆ address global grand challenges through international collaborations

☆ build research and education capacity in both developed and developing countries,

☆ extend communication networks to facilitate virtual research experimentation and data sharing more broadly among the world's science and technology research communities,

☆ Inform policy makers through high-quality, interdisciplinary research.

☆ The Fulbright programme is an opportunity for Zimbabweans collaborate with their U.S. counterparts on curriculum or faculty development projects, or to invite an American expert to be a resource person at a seminar or workshop

☆ KOICA/JICA/Presidential scholarship programmes- also offer the same opportunities as the Fulbright programme

☆ Trans-boundary issues and challenges or Governance of international spaces- International spaces beyond national jurisdictions – including, the high seas, the deep sea and outer space – cannot be managed through conventional models of governance and diplomacy, and will require flexible approaches to international cooperation, informed by scientific evidence and underpinned by practical scientific partnerships. These issues not only present unique foreign policy challenges because of their proximate nature, but, given the strong domestic components, they have active and vocal domestic constituencies. These issues are often set in the context of the natural world, as is reflected in the adage "nature knows no boundaries," Science diplomacy is one of the most promising areas of innovation for how to deal with the great transnational challenges of this century, including nuclear disarmament, climate change, food security, disease, and many other aspects of international peace building{8}

☆ In a speech at the 2008 Davos World Economic Forum, Microsoft Chairman Bill Gates, called for a new form of capitalism, that goes beyond traditional philanthropy and government aid. Citing examples ranging from the development of software for people who cannot read to developing vaccines at a price that Africans can afford, Gates noted that such projects ".provide a hint of what we can accomplish if people who are **experts on needs** in the developing world meet with **scientists who understand** what the breakthroughs are, whether it's in software or drugs." He suggested that we need to develop a new business model that would allow a combination of the motivation to help humanity and the profit motive to drive development. He called it "creative capitalism," capitalism leavened by a pinch of idealism and altruistic desire to better the lot of others. {4}

☆ Scientists and engineers have an important role to play in creating what New York Times columnist Tom Friedman calls a **"flat world,"** a world of economic opportunity made equal through electronic communication technologies.

☆ *Science and Diplomacy* can be a resource for foreign policy makers and analysts, scientists and research administrators, and educators and students in their efforts to better bridge science and foreign affairs. Our goal is a foreign policy that can fully address the increasingly complex technical dimensions of twenty-first century international relations.{2}

## 7. Challenges

☆ Regulatory barriers, such as visa restrictions and security controls, can also be a practical constraint to science diplomacy *e.g.* stringent visa requirements limit travelling opportunities for scientists and scholars

☆ In Zimbabwe and other developing there is still need for a clearly defined system of innovation which shows linkages and a well-coordinated structure of dealing with science and technology research and development. Absents of such a structure, affects and/or delays decision making hence progress is affected and development is hindered

☆ Sanctions-smart/economic sanctions also impact negatively on diplomatic relations across all sectors.

☆ Lack of financial resources- the dwindling government coffers have impacted negatively on all sectors of the economy especially projects which would require long periods of time and resources for them to bring results. Government support on science research and development has been close to none, leaving researchers to source funds through their own networks. Lack of treasury support divorces the researcher from the diplomats. This then reduces the chances of science influencing foreign policy.

☆ Political influence. Many developing countries have unpredictable/harsh political climates and this impact negatively in trying to pursue any form of diplomacy. Unstable political environment chases away potential investors.

☆ Brain drain-as most of the knowledgeable human resources is going for greener pastures, they leave vacuums in their own countries (Zimbabwe is no exception). These vacuums means less expertise to take charge of the science diplomacy process

☆ Fear of technology- some of our political leaders fear science to the extent they do not support research or absorption of emerging technologies as and when they come. If this is the case it means diplomacy in science related matters legs behind.

☆ Misinformation-most policy makers or diplomats are misinformed or lack the correct knowledge in science and technology such that they would not support it.(issues relating to biotechnology/GMOs or use of ICTs)

☆ Inadequate science and technology infrastructure

☆ Science and Technology issues are largely alien to, and almost invisible within most multilateral institutions. Science and Technology, on one hand, and international policy, on the other, are effectively two solitudes, existing in separate, floating worlds which rarely intersect. When diplomats or politicians talk about international policy, you rarely hear anything about S&T, and vice versa.

## 8. Recommendations

☆ Science diplomacy needs support and encouragement at all levels of the science community. Younger scientists need opportunities and career incentives to engage with policy debates from the earliest stage of their careers. There is much to learn from related debates over science communication and public engagement by scientists, where there has been a culture change within science over the past ten years.

☆ Establishment of a scientific liaison desk at all embassies so as to have an in-depth understanding of the policies, people and priorities of their host nation and create opportunities for scientists, universities and high-tech firms at home,

☆ Most developing countries lack science culture-there is need to demystify science and make it an integral part of every individual in the community. In pursuit of a knowledge economy; all communities must be scientifically empowered. A lot of awareness is needed on the importance of science and technology in communities

☆ Develop a science diplomacy policy-whose main objectives should be:

☆ negotiating the participation of scientists in international research programmes;

☆ providing scientific advice to international policymaking;

☆ helping to build science capacity in developing countries and

☆ Using science to project power on the international stage, in ways that increase prestige and attract inward investment.

## 9. Conclusion

Together, science and diplomacy have enabled the human race to delve deeply into the nature of the world around us, to reach across borders and nationalities, and perhaps most importantly to delve deeply into ourselves. Science and technology diplomacy is very vital because it liberates scientific and technological (S&T) knowledge from its rigid national and institutional enclosures and to unleash its progressive potential through collaboration and sharing with interested partners' world-wide. Through science diplomacy, we can overcome any obstacle, bridge any chasm, and solve any problem. Through science diplomacy, we can take the next step forward in our own betterment. Through science diplomacy, we can truly make a difference.

## REFERENCES

1. http://www.ijhssnet.com/journals/Vol_2_No_2_Special_Issue_ January_2012/30.pdf

2. Turekian, Vaughan C.; Neureiter, Norman P. (9 March 2012). "Science and Diplomacy: The Past as Prologue". Science and Diplomacy

3. The First Pugwash Conference". Pugwash Conferences on Science and World Affairs. Retrieved 17 July 2012.

4. Gates, Bill (24 January 2008). "Bill Gates: World Economic Forum 2008 - Transcript of remarks by Bill gates at World Economic Forum 2008"

5. Scientific liaison Office (1983) Cabinet Office Zimbabwe, Research Index, 1981-1983, Harare, Zimbabwe

6. Science and Innovation Policy: Science diplomacy". SciDev.Net

7. Sibanda, F.(2011) African Blitzkriegin Zimbabwe: Phenomenological Reflections on Shona Beliefs on Lightning. Saarbrucken: Lambert Academic Publishing GmbH and Co. KG.

8. http://www.aaas.org/sites/default/files/New_Frontiers.pdf

9. http://worldstemworks.org/about-us/what-is-science-diplomacy/

10. http://www.chiefscie.for-solutions/]

11. GO SPIN Zimbabwe country profile(in the making)

12. http://harare.usembassy.gov/fulbright.html

# — *Section II* —
# Regional Cooperation and South-South Relations

*Chapter 7*

# Better Diplomacy and Better Science for Better Development: A Way Forward Fulfilling Post-2015 Development Agenda and Sustainable Development Goals

*Ruckmani Arunachalam[1], Rita Gupta[2] and Sadhana Relia[3]*

*International Multilateral and Regional Cooperation Division,*
*Department of Science and Technology,*
*Technology Bhavan, New Mehrauli Road, New Delhi-110016*
*e-mail: [1]ruckmani.a@nic.in, [2]ritagupta@nic.in, [3]srelia@nic.in*

## ABSTRACT

Science, as a field is universal and borderless in nature, and can be instrumental in stimulating constructive dialogue, openness and mutual respect among countries. Promoting scientific approaches in public is essential to maintain peaceful democratic societies, where debates build on shared knowledge and where fact-based policies facilitate the emergence of mutually acceptable solutions. This is also true at the international level, where peace and reconciliation can be achieved through scientific cooperation between scientists living in regions marred by conflict.

Emerging and fast-evolving challenges call for innovative solutions for global sustainability. Keeping in mind the borderless nature of the global environmental crises, especially pertaining to climate change and biodiversity, international cooperation will have to gain momentum including intensified S&T collaborations and increased technology transfers.

UN recognises the critical role and contribution of science, technology and innovation in building and maintaining national competitiveness in the global economy, addressing global challenges and realizing sustainable development. It also recognizes the instrumental role of STI, and particularly ICTs, in the achievement of a number of the Millennium Development Goals (MDGs), and the need to make STI figure more prominently in the Post-2015 Development Agenda and Sustainable Development Goals

United Nations Commission on Science and Technology for Development seeks international cooperation opportunities in ICTs, particularly in terms of identifying best practices on e-education, e-government, e-health and disaster resilience through existing and new cooperation platforms. It identifies a need for sovereign wealth funds, for science, technology, engineering and innovation-based solutions to address infrastructural needs for sustainable development, in the post-2015 developmental framework.

UN stresses the need to nurture joint collaborations for capacity-building of human resources and global research infrastructure in scientific, technological, engineering disciplines. At the context, therefore, science diplomacy will serves as a platform for sharing examples of good practice and promoting North-South and South-South partnerships, especially in regard to new and emerging technologies.

By sharing our best practices and successful innovation models, India can also offer novel possibilities for developing countries and help them to bridge the social gap and digital divide. New approaches to development should be meaningful locally and effective globally.

This paper shares some Indian innovation models that had helped many of the developing countries so far in their development. It also discussed Indian approaches for sustainable development and inclusive innovational developmental agenda, along with road maps for effectively utilizing science diplomacy.

*Keywords: India, S&T diplomacy, Sustainable development goals, STI policy.*

# 1. Introduction

Science diplomacy is the use of scientific collaborations among nations to address common problems and to build constructive international partnerships. Science diplomacy has become an umbrella term to describe a number of formal or informal technical, research-based, academic or engineering exchanges. Quoting from American Association for Advancement of Science, the relationship between science and diplomacy can be articulated as three concepts

☆ Science in diplomacy": Science can provide advice to inform and support foreign policy objectives.

☆ "Diplomacy for science": Diplomacy can facilitate international scientific cooperation.

☆ "Science for diplomacy": Scientific cooperation can improve international relations. Any particular international science cooperation activity such as capacity building, joint research projects, fellowships, exchange visits can be described by one or more of these concepts.

# 2. Role and Importance of Science Diplomacy

In an increasingly knowledge driven global economy, S&T competitiveness is critical for long term sustainable development of any country. The role of governments is changing as the conventional economies are undergoing a fundamental transformation to knowledge based economies. Countries have to respond with policies, programs, institutions and partnerships that will maximize their economic opportunities and sustain the social fabric. STI also offers a unique opportunity for exploiting development in untapped areas. Science as a tool for diplomacy has been used for several decades by many countries around the world. One of the earliest ventures in joint scientific cooperation was in 1931 with the creation of the International Council of Scientific Unions, now the International Council of Science (ICSU) which focuses resources and tools for further development of scientific solutions to the world's challenges such as climate change, sustainable development and polar research, and the universality of science. Major challenges facing the world today are ensuring food security, supplying clean water, battling infectious diseases, mitigating climate change, addressing urbanization, building green energy economies and reducing biodiversity loss-all these require transformational and innovative solutions.

Some examples of Science Diplomacy efforts from developed countries include:

Science Diplomacy initiatives have figured prominently in the actions of the developed world especially in the last decade. To illustrate,

☆ Swiss State Secretariat for Education Research and innovation, in cooperation with Swiss Federal Department of Foreign Affairs, set up 'Swissnex' - a network with nodes in world's most innovative hubs: Boston, San Francisco, Singapore, China and India, during 2000-11. This network, in addition to science sections at Swiss embassies abroad, takes an active role in strengthening Switzerland's leadership as a world class location for science, technology and innovation.

☆ Japan's approach towards promoting Science Diplomacy is reflected in their 2008 document: "Toward the Reinforcement of Science and Technology Diplomacy" issued by Council for Science and Technology Policy, Ministry of Education, Science and Technology (MEST).

☆ French Ministry of Foreign Affairs (Directorate-General of Global Affairs, Development and Partnerships) has issued a report on 'Science Diplomacy for France' in 2013.

☆ International and National Science Academies also gave a push to nurturing Science Diplomacy. Noteworthy amongst these are UK's Royal Society which brought out a document, 'New Frontiers in Science Diplomacy' in 2010; USA's American Association for Advancement of Science (AAAS) which has set up a 'Centre for Science Diplomacy 'in 2008; TWAS–AAAS Agreement in 2011 which has been instrumental in bringing out a quarterly publication by AAAS (Science and Diplomacy) from March 2012 onwards.

☆ UNCTAD (Division on Investment, Technology and Enterprise Development) in 2003 brought out a publication on 'Science and Technology Diplomacy: Concepts and Elements of a Work Programme'.

## India's International S&T Cooperation

Department of Science and Technology (GOI) deals with the International Scientific and Technological Affairs including negotiations and implementation of Scientific and Technological Cooperation/Agreements and takes up the responsibility for scientific and technological activities of international organizations. These co-operations are sought under bilateral, multilateral or regional framework modes for facilitating and strengthening interactions among governments, academia and industries in areas of mutual interest. The Department operates in close association with line ministries, MEA, Indian missions abroad, and Foreign missions in India and UN bodies. Science Counselors are also posted in Berlin, Moscow, Tokyo and Washington to facilitate cooperation with respective countries to which they are accredited. India currently has bilateral S&T cooperation agreements with the aim of fulfilling Capacity Building in S&T; Improving policy enabling environment; Human Resource development; Institutional development; Science, Technology and Innovation for development; Knowledge Transfer and Adoption; Identification of Common Research priority areas.

India's exchange of knowledge and experience with other developing countries is driven by perceived need in development partners thus resulting in research cooperation in technologies and mobilizing financial and human resource development.

## Science Diplomacy Efforts in India

India's International partnerships in Science and Technology has benefitted India in many ways. Engagements clearly attributable to 'Science Diplomacy' among others include the following:

☆ Financial and scientific contributions to sustain international S&T organizations such as The World Academy of Sciences (TWAS) and Centre for Science and Technology of the Non-Aligned and Other Developing Countries(NAM S&T Centre)

☆ Contributions to maintain the level and quality of collaboration that can be supported out of 'Commonwealth Fund for Technical Cooperation (CFTC),

☆ Creation of SAARC Development Fund and SAARC Regional Centre on Disaster Management hosted by India ;

☆ Implementation of 'New Africa Initiatives in S&T' with dedicated GOI funds;

☆ Training of developing country scientists in India out of GOI's Indian Technical and Economic Cooperation (ITEC) scheme ;

☆ Hosting developing country scientists for specialized training at centres like Centre for Space Science and Technology Education in Asia and Pacific (CSSTE-AP), Dehradun, National Institute of Hydrology (NIH) Roorkee, Central Food Technological Research Institute (CFTRI) Mysore, Indian National Centre for Ocean Information Services (INCOIS) Hyderabad, Vector Control Research Centre (VCRC) Pondicherry etc.;

☆ Implementation of Research Training Fellowships for Developing Country Scientists (RTF-DCS) programme to promote India as preferred scientific destination through NAM S&T Centre with funding by GOI;

☆ Initiative for ASEAN Integration serving the needs of CLMV nations;

☆ Setting up of a 'Regional Centre for Biotechnology Training and Education in India' in collaboration with UNESCO ;

☆ Setting up of Bay of Bengal Initiative for Multi-Sectoral Technical and Economic Cooperation (BIMSTEC) Centre for Weather and Climate (BCWC) in India to enable all BIMSTEC countries to pool their scientific resources for providing weather and climate related information and services;

☆ Supporting establishment of Rajiv Gandhi Science Centre and setting up of Radio Telescope facility in Mauritius.

☆ Actively participating in various international treaties and protocols/ membership in inter-governmental bodies; European Organization for Nuclear Research (CERN), Antarctica Treaty, Kyoto Protocol on Greenhouse Gas emission reduction, Montreal Protocol on Ozone layer depletion, United Nations Framework Convention on Climate Change (UNFCCC), Inter-governmental Panel on Climate Change (IPCC), World Meteorological Organization (WMO), etc.,

☆ Conducting professional courses for foreign diplomats as well as for Indian Foreign Service probationers by Foreign Service Institute (FSI) of Ministry of External Affairs (MEA);

☆ Setting up of bi-national S&T Centres such as Indo-French Centre for the Promotion of Advanced Research (IFCPAR); The Indo-US Science and Technology Forum(IUSSTF);Indo-German Science and Technology Centre (IGSTC); Indo Russian Centre for Science and Technology (IRCST) with the aim to

   ❐ advance industrial research partnership with mutuality of interest and respect

   ❐ create platform for cross fertilization of ideas

   ❐ develop knowledge networks for industrial sectors to enhance competitiveness

   ❐ establish joint knowledge pools to address global challenges

☆ India's contribution to Global Fund to fight AIDS, Tuberculosis and Malaria.

## Indian Development Cooperation (DC): India's Comprehensive Vision in Helping Developing Countries

India's involvement in DC is not new. It has, since independence, historically supported countries in the South, primarily in the South Asian region, through various forms of assistance. Indian development cooperation is different from other emerging powers, it is complementary to Chinese and traditional aid. In 2003

GOI decided to discontinue the practice of extending loans or credit lines to fellow developing countries, instead to provide grants or project assistance to developing countries in Africa, South Asia and other parts of the developing world viewing them as development partners and thus announced India Development Initiative to leverage and promote our strengths abroad. It can be stressed here that as science and technology occupies the main stream of development both inside and outside our country. Devising of new ways of development cooperation through S&T is the need of the hour. For example India's development cooperation with Africa expanded significantly in 2005 when India became the first Asian country to become a full member of the African Capacity Building Foundation (ACBF) which is based in Harare. The ACBF has emerged as one of the premier organizations for sustainable development and poverty alleviation for Africa. India and regions in the African continent are facing certain common challenges, since the agro climatic conditions are similar in the two regions with common challenges confronting the agricultural sector, Africa and India can opt for similar approaches in addressing them. Theses common challenges and trajectories give huge complementarities for tapping the Africa India cooperation in various fields for finding out common solutions. This also helps in institutionalizing STI. The Africa-India STI cooperation offers opportunity for agricultural growth through value addition to agricultural produce. This may also need bilateral cooperation for facilitating the application of scientific knowledge along with promoting the entrepreneurial activities in the two regions. In this connection under the Africa-India Science and Technology Initiative, DST in partnership with Ministry of External Affairs (MEA) broadly outlined the contours of Africa-India collaborative engagements in consultation with the African Union. In keeping with its growing stature in international affairs, India is willing to assume greater responsibility in promoting development through its various Ministries and Agencies.

## Establishment of Common Platform for Addressing the Post-2015 Development Agenda and Sustainable Development Goals under South-South Cooperation (SSC)

UN recognizes the critical role and contribution of science, technology and innovation in building and maintaining national competitive in the global economy, addressing global challenges and realizing sustainable development. It also recognizes the instrumental role of STI and particularly ICTs in the achievement of a number of Millennium Development Goals (MDGs) and the need to make STI figure more prominently in the Post-2015 Development Agenda and Sustainable Development Goals. A high level panel set up by the UN to provide guidance and recommendations building on MDGs identified five priority transformations for the post-2015 development agenda. These are i) no one left behind; ii) sustainable development at the core; iii) economic transformation for job and inclusive growth; iv) peace and effective open and accountable institutions for all and v) a renewed global partnerships (May 2012). The panel believes the effective implementation of these 5 transformative shifts can end poverty and inequality and promote inclusive and sustainable development. As a way forward the panel has proposed the following Sustainable Development Goals to be attained by 2030:

1. End poverty everywhere
2. End hunger, improve nutrition and promote sustainable agriculture
3. Attain healthy lives for all
4. Provide quality education and life-long learning opportunities for all
5. Attain gender equality, empower women and girls everywhere
6. Ensure availability and sustainable use of water and sanitation for all
7. Ensure sustainable energy for all
8. Promote sustained, inclusive and sustainable economic growth, full and productive employment and decent work for all
9. Promote sustainable infrastructure and industrialization and foster innovation
10. Reduce inequality within and between countries
11. Make cities and human settlements inclusive, safe and sustainable
12. Promote sustainable consumption and production patterns
13. Tackle climate change and its impacts
14. Conserve and promote sustainable use of oceans, seas and marine resources
15. Protect and promote sustainable use of terrestrial ecosystems, halt desertification, land degradation and biodiversity loss
16. Achieve peaceful and inclusive societies, access to justice for all
17. Strengthen global partnership for sustainable development.

Only STI can lead to affordable access of quality goods and services creating livelihood opportunities for the excluded population, primarily at the base of the pyramid and on a long term sustainable basis with the significant outreach. The understanding of the fundamental principles and modalities of SSC will help foster common ground among the South in preparation for the Post-2015 Development Agenda and Sustainable Development Goals. SSC is built on its demand driven approach; non-conditionality; respect for national sovereignty, national ownership and independence and above all mutual benefit. It promotes self-reliance by offering opportunities for development partners to pursue collectively the goal of sustainable development. The modalities of SSC have taken different forms which include capacity building, training, technology transfer, and financial assistance. The diversity of SSC is enriching and valuable. The changing development landscape and global norm-setting increasingly calls on Southern partners to coordinate on strategy, policy and operations, especially in fields like agriculture and health care. The Millennium Development Goals are widely recognized for having focused the development agenda on a set of clear and reachable goals for galvanizing global efforts to achieve them. Now we have post-2015 SDGs before us. As rightly pointed out by the Secretary-General of UNCTAD in his speech during Open Working Group on SDGs, the critical role of advancing technological progress and promoting structural transformation as foundation of sustainable development and the need to match the scope and ambition of goals and targets in terms of resources and policy reforms. Health related SSC can be a means towards achieving the Millennium Development Goals, such as reducing child mortality, improving maternal health, combating HIV/AIDS, malaria and other diseases (Chaturvedi, 2013).

Emerging economies like Brazil, India, China and South Africa and other countries such as Cuba have come forward to supplement global efforts in tackling various health-sector related challenges. Common interest from shared health concerns in developing countries is the key driver for India's South-South collaborations. Malaria is one disease which is now essentially limited to tropical countries. Leprosy and cholera are still a problem in many countries in South–East Asia, Africa and Latin America. South-South partnerships will strength advocacy and help combine health initiatives while making more resources available for the fight against infectious diseases. Eight manufacturers in India currently produce 60-80 percent of all vaccines produced by UN agencies making India by far the largest provider of affordable, high-quality vaccines for developing countries. Energy is a field where scientific cooperation will be beneficial to all countries especially in South Asian region for generation, transmission and efficient use. Nuclear cooperation in South Asia can include sharing of nuclear power across national borders and in other spin-off areas like health, agriculture and hydrology etc., Therefore science diplomacy should lead to Synergy in Energy and other S&T technologies under S-S cooperation. Energy security is the basis of sustainable development.

## India's STI Policy 2013 Addressing Sustainable Development Goals

According to Brundtland Commission Report (1997) Sustainable Development is that "meets the needs of the present without compromising the ability of future generations to meet their own needs". For sustainable development to take place conduce policy framework is necessary. Science, Technology and Innovation (STI) offers unique opportunity for exploiting development opportunities in untapped areas that can be critical to empowering excluded populations. STI have emerged as the major drivers of national development globally. As India aspires for faster, sustainable and inclusive growth, the Indian STI system with the advantages of large demographic dividend and huge talent pool will need to play a defining role in achieving these national goals. India's STI policy takes cognizance of United Nations Millennium Development Goals and post-2015 sustainable development agenda. The pressing problems most of the developing countries in general are:

☆ Lack of progress in attaining food security (Goal 2)
☆ Energy independence (Goal 7)
☆ Efficient water management (Goal 6)
☆ Tackling climate change (Goal 15)
☆ Providing universal health care (Goal 3)

These problems can only to be addressed by interdisciplinary solutions with use of science and technology. Overall, the guiding vision for new Indian STI policy 2013 is faster, sustainable and inclusive growth. The inclusive innovation and inclusive economic growth will eliminate the social disharmony (due to poverty, distance, disability and migration) which is also one of the post-2015 sustainable development goals of UN. The key challenge is "Access Equality" despite "Income in equality". Over the past half century, significant advances in international development such as antibiotics, vaccines, cell phones, and mobile technologies have dramatically changed the trajectory of developing countries for the better.

India is working in low cost interventions in prevention, screening and management of diabetes, low cost community interventions to reduce burden of diseases and its complications and makes sure that affordable health care is available to remote areas and would like to have collaborations with regard to use of technology, diagnostics for screening and monitoring. India's novel, innovative national STI programs include a) Innovation in Science Pursuit for Inspired Research-INSPIRE; b) fellowships for re-entry of women who had break in careers due to social and other reasons; c) Consolidation of University Research, Innovation and Excellence (CURIE) for women only universities; d) Power of Ideas program to encourage creative minds into innovative products-services and providing mentoring and financial back-up support. The CSIR (GOI) led Open Source Drug Discovery initiative is a Team India Consortium with global partnerships working towards affordable health care for all. With over 7000 members from more than 130 countries the OSDD project was launched in September 2008 for Tuberculosis as first disease target and is now being extended to Malaria. OSDD works as a distributed virtual laboratory with collaborations from across the globe. Besides, CSIR-800 is scheme of S&T intervention for improving life style and quality of lives of our population in rural India. India aims to working on increasing production of low cost high quality foods to improve diet quality among the poor. Novel Therapeutic Food for Management of Severe Acute Malnutrition (SAM) have been developed in India under Department of Biotechnology (DBT) funded projects as per WHO specifications and tested. These have specific compositions and were proved to be effective for functional recovery. ULTRA RICE®- UR is a micronutrient-fortified, manufactured food product that was developed in India to address nutritional deficiencies in populations where rice is a staple food. Now DBT would be providing the necessary technical know-how for setting up production facilities in States(s); State Health Departments.

Earlier GOI has declared 2010-20 as the "Decade of Innovation" and also established the National Innovation Council. The STI Policy 2013 is in furtherance of these pronouncements. Science, Technology and Innovation in its integrated form will lead to new value creation and India's global competitiveness will be determined by the extent to which STI contributes for social good and economic wealth. Therefore STI for people is the new paradigm of the Indian STI enterprise. Innovation for inclusive growth implies ensuring access, availability and affordability of solutions to a large population as possible. The instruments of new policy 2013 will enable this to be realized in India. Emphasis will be to bridge the gap between the STI system and socio-economic sectors by developing a symbiotic relationship with economic and other policies. The policy promises to focus to prioritizing critical R&D areas like agriculture, telecommunications, energy, water, management, health and drug discovery, environment and climate change. The policy also seeks to establish a new regulatory framework for data access and sharing as also for creation and sharing of IPRs.

The key elements of the STI policy include:

☆ Enhancing skill for applications of science among young from all social strata

☆ Establishing world class infrastructure for R&D for gaining global leadership in some frontier areas of science.

☆ Linking contributions of science, research and innovation system with the inclusive economic growth agenda and combining priorities of excellence and relevance.

☆ Enabling conversion of R&D outputs into societal and commercial applications by PPP structures.

STI policy envisages promoting excellence and relevance in R&D by:

☆ Nourishing the roots

☆ Ensuring Gender Parity

☆ Participation in Global R&D infrastructure

☆ Attracting private sector investments in R&D

☆ Fostering mobility of experts from academia to industry and vice-versa.

## India – as a Global Leader in Science: Looking Forward

The way forward to bring post -2015 sustainable development goals has to be built upon current S&T initiatives, and by identifying new ones, building synergies and in general bringing all the stakeholders-government, policy makers, public sector, private sector, academic agencies and research organizations. Science and Technology need to be brought into close association with Indian business and government communities to jointly examine the use of S&T under development cooperation. Science Diplomacy, as a distinct discipline merits attention and concrete actions through N-S and S-S partnerships under the overall ambit of international relations. The actions dedicated to Science Diplomacy could include proactive initiatives by Government(s) in: bringing out white papers, policies and case studies; conducting training programmes and workshops; introducing specialised post graduate diploma courses; expanding doctoral research programmes; science diplomacy platform for networking of institutions, science academies and experts. Institutional structures in Science Diplomacy in the form of Science Consulates, Science Counselors for African Union, ASEAN, China and European Union are also desirable. India could take a lead in augmenting institutional and human capacity building of developing countries in science diplomacy by establishing a "Centre for Science Diplomacy" at an appropriate location within India with an aim to promote the use of science in building bridges between nations in South-South and North-South cooperation modes and raise the status of scientific cooperation to being a vital element of foreign policy of any country. The focus will be on achieving economic growth and prosperity of a country by analyzing appropriate modes of bilateral, multilateral or regional cooperation arrangements and executing them through science diplomacy strategies.

## 3. Conclusion

In conclusion, science, if perceived as a global public good and not as a national treasure can play an important role in defining common action in regard to the Sustainable Development Goals. Strong national ownership, well-managed policies, peaceful socio-political and stable economic environments supported coherently by all partners including the UN systems critical for global success of sustainable development goals. It is very much appropriate to end with the following quote from our first Prime Minister Pundit Jawaharlal Nehru:

*"The service of India means the service of millions who suffer. It means the ending of poverty and ignorance and disease and inequality of opportunity. The ambition of greatest men of our generation has been to wipe every tear and suffering, so long our work will not be over. And so we have to labour and to work…to give reality to our dreams. Those dreams are for India but they are also for the world, for all the nations and people are too closely knit together today for any one of them to imagine that it can live apart. Peace is said to be indivisible, so is freedom, so is prosperity now and also disaster in this one world that no longer be split in to isolated fragments."* (*Government of India 2011*).

## 4. Acknowledgements

The authors thank Secretary, Department of Science and Technology, New Delhi, India for giving permission to present this paper at International Workshop on Perspectives on Science and Technology Diplomacy for Sustainable Development in NAM and other Developing Countries held at Heritage Village, Manesar, Haryana during 27-30th May 2014. Authors also acknowledge Dr.N.G. Satish, Librarian, ASCI, Hyderabad, India and Dr. Sachin Chaturvedi, DG,RIS, New Delhi for providing useful reference materials.

## 5. Abbreviations

|        |                                                        |
|--------|--------------------------------------------------------|
| ASCI:  | Administrative Staff College of India                  |
| ASEAN: | Association of South East Asian Nations                |
| CSIR:  | Council of Scientific and Industrial Research, Govt. of India. |
| DBT:   | Department of Biotechnology                            |
| GOI:   | Government of India                                    |
| ICT:   | Information and Communication Technology               |
| MDGs:  | Millennium Development Goals                           |
| OSDD:  | Open Source Drug Discovery                            |
| RIS:   | Research Information Systems                           |
| SDGs:  | Sustainable Development Goals                          |
| SSC:   | South-South Cooperation                               |
| S&T:   | Science and Technology                                |

STI: Science, technology and innovation

UN: United Nations

UNCTAD:United Nations Conference on Trade and Development

WHO: World Health Organization

## REFERENCES

1.  Alex Dehgan and E. William Colglazier, "Development Science and Science Diplomacy," *Science and Diplomacy*, Vol. 1, No. 4 (December 2012).

2.  Meeting of UN System Task Team on the Post-2015 UN Development Agenda, May 2012.

3.  Science, Technology and Innovation Policy 2013, Ministry of Science and Technology, New Delhi, Government of India.

4.  Science, technology and innovation for the post-2015 development agenda-Report of the Secretary-General at the 17th session of Commission of Science and Technology for Development at Geneva, during 12-16, May 2014 (E/CN.16/2014/2)

5.  New frontiers in Science diplomacy: Navigating the changing balance of power, The Royal Society –London Jan, 2010

6.  Chaturvedi.S and Ravi Srinivas, K 2014, India-Africa Cooperation in Agriculture Science, Technology and Innovation: New Avenues and Opportunities-Forum for Indian Development Cooperation (FIDC) policy brief No:2.

7.  South –South Cooperation: Issues and Emerging Challenges; RIS Conference Report 15-6, April 2013, New Delhi.

8.  Report of the Global Health Strategies Initiative (GHSi), 2012 pg: 50-53.

9.  Science and Technology Diplomacy: Concepts and Elements of a work programme. (UNCTAD/ITE/TEB/Misc.5), New York and Geneva 2003.

10. Report of the Secretary-General at the Substantive session of 2014 of UN Economic and Social Council, New York, 7-11 July 2014 (E/2014/61).

11. Schlegel, F, "Swiss Science Diplomacy: Harnessing the Inventiveness and Excellence of the Private and Public Sectors," Science and Diplomacy, Vol: 3(1) March 2014.

12. Government of India (2011) Inaugural address delivered by the external affairs minister of India on 'Harnessing the positive contribution of South-South Cooperation for development of least developed countries (LDCs) at the India-Least Developed Countries Ministerial Conference, New Delhi, 18 February.

13. Background papers for the FIDC conference on Indian Development Cooperation Policy: The State of the Debate; 18 Jan 2014.

14. Toward the Reinforcement of Science and Technology Diplomacy (Provisional Translation) May 2008, Council for Science and Technology Policy, Japan.

15. Chaturvedi Sachin (2013), External Health Aid and Sustainable Development: Emerging contours of Indian Health Diplomacy", Presentation made at Regional World Health Summit, Singapore.

16. Strategic report on "Science Diplomacy for France." by The French Ministry of Foreign Affairs (Directorate-General of Global Affairs, Development and Partnerships - Mobility and Attractiveness Policy Directorate) 2013.

17. www.dst.gov.in

*Chapter 8*

# Enhancing Technical and Vocational Education through Science and Technology Diplomacy

*Aworanti Olatunde Awotokun*

*Registrar/Chief Executive,*
*National Business and Technical Examinations Board,*
*Nigeria*
*e-mail: aworantio@yahoo.com*

## ABSTRACT

Science and technology diplomacy involves those international and national science and technology activities directed towards international commitments to global development in general and the third world countries in particular. The diplomacy entails governmental negotiations in which specialization and integration are handled. In the light of this, apt negotiations have been made by Nigerian government with the western world in the area of providing support for technical and vocational education and training (TVET) as well as in industrial development. In pursuit of multilateral relations, many international organisations still embark on supporting and partnering with Nigerian government in knowledge and skills development areas. This paper is therefore prepared to evaluate the roles of science and technology diplomacy in enhancing technical and vocational education and training in Nigeria. Efforts are equally made to appraise the best practices in science and technology diplomacy which stand as benchmarks for developing nations, Nigeria inclusive. The appraisal however revealed that Nigeria would advance technologically if science and technology diplomacy could be employed in the areas of training of technocrats, science and technology partnership, technology sourcing, financing and assessment and evaluation in TVET system while the political will should be made TVET-centred.

*Keywords: Science, Technology, Diplomacy, Science and technology diplomacy, TVET.*

# 1. Introduction

In the first decade of the advent of formal education in Nigeria, there was no conscious effort at directing such system of education to national development as noted by Nduka (1965) and Fafunwa (1982). The duo viewed this type of education as parochial and irresponsive to the peculiar needs and aspirations of the evolving Nigeria. The search for a more responsive curriculum for education in Nigeria led to the national conference of 1969 and the report of committee set up in 1973 resulted in the preparation of the white paper titled "National Policy on Education" in 1977 (Ikuenobe, 2014).

The formal education system 6-3-3-4 evolved in 1991 with a widened curriculum aimed at equipping the Nigeria child with adequate intellectual capacity with respect to dignity of labour. This new structure provided variety, both academic and technical at various levels throughout the education system and placed emphasis on Technical and Vocational Education and Training (TVET).

Okafor (1992) and Toby (1997) stressed the importance of TVET for economic development and individuals as the channel through which people are prepared for occupations requiring manipulative skills.

TVET, according to the 1977 UNESCO International Standard Classification of Education, is the type of education and training leading to the acquisition of practical skills, technical knowhow and understanding required for employment in a particular occupation, trade or group of occupations or trades. It equips the beneficiary with overall economic development of the society (Afeti 2010). The overall objective of TVET is to provide the economy with qualified and competitive workers and to train citizens for sustainable growth and poverty reduction by offering training opportunities to all social groups without discrimination.

Technical and Vocational Education and Training system is designed to:

☆ Assure guidance and counselling, planning, coordination, monitoring and evaluation of TVET activities.

☆ Provide theoretical and practical trainings, capable of meeting the needs of enterprise in all sectors and sustaining international standards.

☆ Satisfy quantitative and qualitative needs of priority sectors by training the much needed manpower in the relevant areas.

☆ Provide the graduates with required skills for professionalism *i.e.* ensure their employability and develop their ability to learn independently during this professional life without any form of discrimination and prepare them for self-development and;

☆ Develop work values and positive individual work attitudes towards professionalism expressed in quality efficiency, creativity, adaptability, commitment, responsibility and accountability, the spirit of service and genuine love of well done work.

TVET is being financed by both government and individual companies for their developmental needs. Science on the other hand, is the concerted human effort to understand or to understand better, the history of the natural world and how it works with observable physical evidence as the basis of that understanding (Gly, 2014). Technology is a Latin word meaning tools, materials and a process of solving practical problems. According to Simiyu (1999), the term technology as applied to the process of education includes ways of organizing events and activities to achieve education objectives as well as the materials and equipment involved in the process. Technology can be described as a production in the sense that it is the end result of the systematic application of scientific knowledge in addressing human learning problem (Adegbija, Fakomogbon, and Daramola, 2012). Defining diplomacy, the advanced learner's Dictionary sees it as the management of relations between countries.

The world has become a global village hence the need for Nigeria to key into science diplomacy to enhance her development. No nation has all the knowledge and technology that are needed for development. Houston Times wrote:

> *"ideas can cross mountains, valleys and seas. They go anywhere a man can go and endure long after he is gone; ideas are indestructible because of their very nature; there is no defense on earth against them"* (Kwakpovwe, 2014).

In 1967, the African Scientific Institute was established to help African Scientists reach others through published materials, conferences, seminars and provide tools for those who lack them (AAAS, 2013).

Diplomacy is the art and practice of conducting negotiations between representatives of states. It usually refers to international relations through the intercession of professional diplomats with regard to issues of peace- making, trade, war economics, culture, environment and human rights. There are a variety of diplomatic categories and diplomatic strategies employed by organisations and governments to achieve their aims. It is a complex and often challenging practice of fostering relationships around the world in order to resolve issues and advance interests.

Science diplomacy is the use of science collaborations among nations to address common problems and to build constructive international partnerships. Many experts and groups use a variety of definitions for science diplomacy which has become an umbrella term to describe a number of formal technical, research – based, academic or engineering exchange (Royal Society, 2014).

In January 2010, the Royal Society and the American Association for the Advancement of Science (AAAS) noted that science diplomacy refers to three types of activities:

☆ "Science in diplomacy" - Science can provide advice to inform and support foreign policy objectives.

☆ "Diplomacy for Science"- Diplomacy can facilitate international scientific cooperation.

☆ "Science for diplomacy" - scientific cooperation can improve international relations.

William (2013) defined science as not only physical and biological sciences, but also the social sciences, engineering and medicine.

He went further to state that

1. Science and technology aiding diplomacy (for the many diplomacy issues where scientific and technological information is critically important and even for those cases where science and technology engagement can open doors for dialogue on other issues);

2. Diplomacy advancing science and technology (such as negotiating multinational arrangement for building large facilities and gaining access for research in unique locations) and

3. Science and technology helping to solve national, regional and global problems (such as creating new options and paths for making progress on the "wicked problems" too difficult for Politicians to resolve alone).

In addition, he stressed that it helps to improve relations between countries and make all people more secure, healthy, peaceful and prosperous.

According to Juma (2013), science and technology are being increasingly recognised as central features in international diplomacy. Much of the attention has focused on how major industrialised countries and large emerging nations such as China, India and Brazil use science and technology to advance their global competitiveness. He stipulated that most pressing global challenge is how to leverage the power of new knowledge to help address the global economic and environmental challenges. However, Technical and Vocational Education and Training (TVET) can be enhanced through science and technology diplomacy. The advantages derived from these interactions by the developing countries can contribute to national development and foster international relationship.

Science and technology diplomacy is the term used for describing scientific and technological collaborations among nations to address common problems and to build constructive international partnerships (Wikipedia, 2014). It is generally used to describe a number of formal or informal technical, research-based, academic or engineering exchanges among nations. These international negotiations or exchanges are basically determined by two key features of the growth of scientific and technological knowledge which include scientific knowledge and application of science and technology to development. Firstly, scientific knowledge is becoming increasingly specialized and therefore demands greater expert input into international negotiations. Secondly, the application of science and technology to development requires the ability to integrate the divergent disciplines that are needed to solve specific problems. International diplomacy now demands that government negotiators deal with both specialization and integration (UNCTAD, 2003).

Historically, United Nations Conference of 1963 on the Application of Science and Technology for the Benefit of Less Developed Areas marked the origins of the United Nations (UN) focus on science and technology as useful and important tools for development. It was precipitated upon the view that the developing countries "leapfrog" from generation to generation by employing technologies developed in the industrialized countries. Different approaches to science and technology advice emerged from the conference among which were inauguration of the Committee on Science and Technology for Development (CSTD) which was to be a new component of ECOSOC, the Advisory Committee on the Application of Science and Technology for Development (ACAST) which comprised experts in the various fields of science and technology who provide science advice to CSTD and other UN bodies and the Office of Science and Technology (OST) which was established as an integral part of the UN Secretariat to support both groups and assist in the implementation of the advice. The advice was however, later faced with serious criticisms.

Following these criticisms, United Nations Conference on Science and Technology for Development (UNCSTD) was held in 1979. Resulting from this conference was the introduction of Vienna Program of Action which necessitated an updated version of the existing mechanism for science and technology development among United Nations and placed emphasis on the need for capacity building and technology transfer. The Intergovernmental Committee on Science and Technology for Development (IGC) was also set up to replace the CSTD in its capacity of setting science and technology directives while the UN Advisory Committee on Science and Technology for Development (ACSTD) replaced the ACAST. These committees consisted of scientists as well as experts from government and business sectors. The OST was equally upgraded and renamed the Centre for Science and Technology for Development (CSTD) while another funding mechanism was put in place by extirpating the UN Financing System for Science and Technology for Development (UNFSSTD).

In 1992, the Commission on Science and technology for Development (CSTD) came into existence following the abolition of IGC and the ACSTD by the General Assembly. The CSTD provides the General Assembly and ECOSOC with high-level advice on relevant science and technology issues through analysis and appropriate policy recommendations or options. Since July 1993, the UNCTAD secretariat has been responsible for the substantive servicing of the Commission. By 2002 the CSTD was mandated to formulate and handle science and technology policies within ECOSOC and the UN at large. The ECOSOC model later became the benchmark for other bodies of the UN in the area of science and technology.

## The Current Status of TVET in Nigeria

According to International Baccalaureate Organisation (2005), every person is expected to have the opportunity to have his or her experiences and skills gained through society or through formal and non-formal training assessed, recognized and certified. Programmes to compensate for skill deficits by individuals through increased access to education and training should be made available as part of the recognition of prior learning programmes. Assessment should identify skill gaps, be transparent, and provide a guide to the learner and training provider. A credible system of certification of skills that are portable and recognised across enterprises, sectors, industries and educational institutions, whether public or private is also expected to be put in place.

In Nigeria, traditional apprenticeship offers the largest opportunity for the acquisition of employable skills in the informal sector whereas formal TVET programmes are school-based. Students into the technical and vocational education track at the end of junior secondary school. Many Nigerians still see technical and vocational education track having unattractive and unenviable reputation of being a dead end so far as academic progression is concerned and fit for those pupils who are not academically sound to proceed to higher education. The flair for TVE on the part of many Nigerians is quite un-comparable with other developed nations of the world. This is illustrated in the Table 8.1.

**Table 8.1: Analysis of Flair for Technical and Vocational Education**

| S/No. | Countries | Proportion of Post-Primary Students that are in TVE Institutions |
|:---:|:---|:---:|
| 1. | United Kingdom | 66 per cent |
| 2. | France | 65 per cent |
| 3. | Germany | 72 per cent |
| 4. | Singapore | 92 per cent |
| 5. | South Korea | Over 50 per cent |
| 6. | Bahrain | Over 55 per cent |
| 7. | Middle East | 50 per cent |
| 8. | Nigeria | 1 per cent |

*Source*: UNESCO (2000).

Technical and Vocational Education and Training (TVET) represents a comprehensive and inclusive approach, intended to help people achieve their full educational and vocational potential and in so doing make meaningful contributions to the communities in which they live. The Federal Republic of Nigeria (2004) defines technical education as that aspect of education which leads to the acquisition of practical and applied skills as well as basic scientific knowledge. Technical education is primarily education for technicians and technologists. It has to do with the development of skills and knowledge to be applied in practical situations.

On the other hand, vocational education is defined as the demonstrated and acknowledged development of knowledge, skills and attitude necessary for a place in the workforce, at levels ranging from pre-trade to prepare professional. (UNESCO,1999). Vocational Education is for pre-professional training or for production of low-level manpower (skilled labour) *i.e.* Artisans, craftsmen and master craftsmen for the labour market. Generally, Vocational Education is education for the craftsmen (Federal Ministry of Education, 2011).

Interestingly, the orientation towards the world of work and development and updating of employable skills curriculum has become a vital characteristics of TVET in Nigeria. The delivery system is designed to train the skilled and entrepreneurial workforce needed for wealth creation and poverty alleviation in the country despite the emerging challenges. TVET institutions also exist to cater for the training needs of learners and prepare them for gainful employment and sustainable livelihoods.

TVE are fully provided in Technical Colleges and Vocational Educational Institutes (VEIs) to those who have acquired basic and post-basic education but could not proceed to the tertiary level. Products or graduates of such institutions are employable in to workforce in Nigeria having acquired the pre-requisite skills in their various trades. Many have become professionals, entrepreneurs and self employed while few others proceed to tertiary institutions for advanced craft/ technical, programmes on their academic progression.

The effort of the formal sector in TVE programmes have been corroborated by the informal sector programmes which are instrumental to the growth and development of Nigeria. This led to such initiatives as public private partnership (PPP) in TVET and the introduction of National Vocational Qualifications Framework (NVQF) which forms the hierarchical system for the development, classification and recognition of generic skills, knowledge and competencies acquired by TVE beneficiaries.

Following the same suit as developed nations that have grossly benefitted from introduction of NVQs, Nigeria is in the process of developing its NVQF to properly assess the non-formal sector of Technical and Vocational Education. This is demonstrated with the introduction of Vocational Enterprise Institutes (VEIs) and Innovative Enterprise Institutes (IEIs) whose products are expected to be awarded National Vocational Certificate (NVC) and National Innovative Diploma (NID) respectively.

## 2. TVET Assessment System

An educative process without assessment is absolutely inconceivable (Nworgu, 2010). According to him, assessment is the pendulum that defines the decision path necessary for the progress of the individual learner and the system at large.

Assessment involves the process of observing, describing, collecting, recording, scoring and interpreting information about a student a group of students (Ekpenyong, 2010). Assessment of any academic programme is viewed to be internal or external. Various educational programmes mounted for the advancement of learning and development of citizenry in view of national development in Nigeria are complimented with proper assessment strategies and exercises of appropriate agencies including examination bodies such as West African Examination Council (WEAC), Joint Admission and Matriculations Board (JAMB); National Business and Technical Examinations Board (NABTEB); National Examination Council (NECO) and other related bodies. Technical and Vocational Education and Training (TVET) has become one of such educational programmes put in place for technological advancement of the nation, Nigeria. Hence the assessment of this system becomes imperative and cannot therefore be overlooked. Assessment of TVET, according to Ekpenyong (2010), not only helps to assess individual student achievement, certify mastery of certain learning competencies and skills, improve instruction and learning, but to assess the effectiveness of vocational curricula, programmes of parts thereof and assess instructional materials.

From the above assertion, it could be deduced that assessing TVET is an all-embracing activity. Thus, it spans through facilities, personnel, curricular contents, instructional methodologies and materials, learners' or trainers' achievement and educational programme evaluation. Such an assessment is also done to ascertain the people/societal interest towards technical and vocational education and training as well as the degree to which TVET has significantly or appreciably contributed to the development of a nation.

## 3.  The Role of Science and Technology Diplomacy in Enhancing TVET in Nigeria

The concept "Science and Technology Diplomacy is about the use of Science and Technology to reach mutual scientific agreement between different nations of the world. Advances in science and technology have become key drivers in international relations and knowledge of trends in key fields is an essential prerequisite to effective international negotiations (UNCTAD, 2003).

Scientific and Technological knowledge is central to Technical Vocational Education and Training (TVET). Science and Technology remain the major tools or subjects for enhancing TVET development at the global level. Therefore, Nigeria is no exception.

The perceived needs for vocational education include the following:

☆ Skills development;

☆ Strategies for addressing unemployment and the social problems of the young and nationalizing qualifications; and

☆ The creation of an alternate route to higher education.

In Nigeria, Vocational Education and Training forms part of secondary education and its programmes prepare learners for the world of work and/or admission to higher education. Different systems of vocational education were introduced in the country namely general vocational and the one named Modular Trades located at workplace. General vocational education has much more in common with general academic than the Modular Trades and would lead to higher education. The learners need to be equipped with foundational knowledge particularly language and mathematical literacy and key social and cognitive skills necessary for effective functioning within modern work situations, including the ability to work in teams, to innovate and to take initiative when appropriate.

The general vocational education programmes will not only prepare learners for specific occupational competence but will also offer them a broad-based orientation to employment, skills as well as sufficient education to prepare them for admission to higher education or the world of work.

## Partnership with Developed Countries in TVET through Science and Technology

The focus of the Science and Technology Diplomacy initiative is to build capacity in developing countries to address more effectively issues related to the role of science and technology in international diplomacy.

Many countries in the world have used technical education to advance their economies to developed ones. Singapore, a former British colony leapt from developing economy to a developed one through the establishment of an Institute of Technical Education (ITE) admitting and training youths who were unable to gain admission into the universities and polytechnics. According to Okafor (2011), Technical Education facilitates the acquisition of practical and applied skills as well as basic scientific knowledge.

Transfer of technology from industrialized to developing countries could enable developing regions to transform their economies at rapid rates. However, there are some limiting factors to the extent to which technology can be transferred to developing countries.

☆ Tariff peaks and escalation with stages of processing.

☆ Lack of capacity to import technology embodied in machinery and equipment, and the vulnerabilities of developing country exports to changes in standards, notably sanitary and phyto-sanitary (United Nation, 2003).

However, Nigeria and other developing countries could attract development in the form of machines and other technical tools for their technical colleges and polytechnics, secondly, in the area of manpower training both within and outside the country. In this way, the much talked about technology transfer would begin to transform the technical and socio-economy environment of the developing nations.

There are various agencies of the United Nation set up to strengthen productive capacity through trade-related industrial and investment measures. These agencies formulate policies that enhance technological development such as 'World Trade Organization (WTO). They set norms for the umbrella body. They are regarded as soft laws, to change behaviour of member states. Various UN – related agencies are also involved in setting technical standards, Examples include the International Civil Aviation Organization (ICAO) and the World Meteorological Organization.

### Best Practices

The importance of science as a subject cannot be over emphasized as it has impacted a lot of values in our lives. These values include; societal values, economic growth and control over our planet and environment. In the case of social values, it has improved peoples' lives by widening their knowledge about genetic information, disease transmission and weather *e.g.* earthquakes, volcanic eruptions and landslides. For economic growth, it has thrown more light on how to recover natural resources by agricultural output, petroleum resources, new chemical substances and technological application and conductivity. In the area of control over planet and environments, it has broadened global knowledge on toxins and waste products especially as it affects water, soil and air. It has also enlightened us concerning causes of climate and their effects on food and water control. It has helped to manage three ecosystems and finally broadened our knowledge of the world because of enlightenment. The idea of science and technology is for sustainable development which can be achieved through the knowledge of the importance of science as mentioned above.

Science has been playing important roles in the lives of individuals both at national and international levels. Royal society (2010) maintained many scientists support international co-operation. This supporting system has made science diplomacy gain ground over the years. Science diplomacy is mainly for science developments that could bring noticeable impacts on national interest. It is used for foreign policy making for the interest of the world.

Science diplomacy is the use of scientific interactions among nations to address the common problems facing humanity and build constructive, knowledge-based international partnership (Royal society, 2010). In recent years, some developed countries like UK, US and Japan have shown great interests in science diplomacy and its practices have made the concept to gain ground. The practices defer, depending on the country's priorities and objectives but generally, these practices among the countries depend largely on the following factors;

- ☆ the mobilization and availability of financial resources
- ☆ external and internal resources
- ☆ partnerships with other institutions (national and international institutions).
- ☆ training programmes
- ☆ sponsorships (national and international levels).
- ☆ networks (private and public)
- ☆ use of expertise on trade, technology, investment and environment.

As earlier said, these practices defer from country to country but with one aim of foreign policy making that will benefit the whole world, which will in turn boost the economic growth of the countries involved.

## US Practice

The Office of Science and Technology adviser to the US Secretary of state was created in 2000. The priorities as sent by the officer in charge (Dr. Nina Federoff) are as follows:

(i) strengthening partnerships across international scientific communities.

(ii) building science capacity within the department of state, and

(iii) horizon scanning for scientific development that could impact on US national interests.

## UK Practice

The United Kingdom (UK) is not left out in the practice. In UK practice, Science and Innovation Network (SIN) was set up in 2001.

In the later year (2009), Professor Daniel FRS was appointed and the Chief Science Adviser to the foreign and commonwealth office. Among their aims is to link science more directly to its foreign policy priorities. In recent years, the practice has expanded to enable a mixture of UK expatriates and locally engaged experts. These are located at UK embassies, high commissions or consulates. These staff work alongside with other diplomats and representatives of countries such as UK, Trade and investment. In terms of research, the network facilitates collaboration between UK and international research partnership across a wide variety of policy and scientific agenda, and the SIN officers are trained to have an in-depth knowledge of the policies, people and priorities of their host countries and finally identify opportunities for UK Scientist, Universities and High-tech forums. This practice can also be adopted by Nigerian government. In this case, Chief Science Adviser can be appointed to represent in Foreign and Commonwealth office. Other trained officers can be appointed in Nigeria embassies for the purpose of Technology and Vocational Education Training (TVET).

## Japan Practice

Japan is another country that practices science and technology diplomacy. There, formal TVET policy on science diplomacy had been in existence since 2007. For a successful practice, four objectives were identified, namely; (1) negotiating the participation of Japanese scientists in international research programmes (2) providing scientific advice to international policy making (3) helping to build science capacity in developing countries, and (4) using science to project power on the international stage. These objectives especially the last one as advised has increased Japan's prestige and attracted inward investment and as such boost their economy.

Nigeria as a developing country is making attempts to catch up with the industrialized economies of the world in the areas of science and technology. In so doing, the Nigerian governments (at all levels) have to pay greater attention to science diplomacy through the teaching of Technical Vocational Education and Training (TVET). Again Nigerian government should embrace the strategies employed by some developed countries in providing science and technology diplomacy.

## Bail Out

To enhance Technical and Vocational Education (TVE) through Science and Technology Diplomacy, the following recommendations are made:

☆ Funding: TVE should be adequately funded to meet international requirement;

☆ Use of Modern Technologies: government should make efforts to partner with industrialized nations in the use of modern technologies in schools;

☆ Public-Private Partnership: for national development, there should be strong collaboration with Public- Private Partnership in African countries to entrench TVE;

☆ Political Will: There should be a strong political will to encourage all sectors of the economy to participate in TVET activities;

☆ Public Enlightenment: There should be aggressive enlightenment for the public to be kept abreast of the benefits derivable from Science and Technology Diplomacy.

## 4. Conclusion

The full participation of Nigeria in TVET with other industrialized nations through science and technology diplomacy will no doubt increase investment in Nigeria which will consequently boost the nation's economy.

## REFERENCES

1. Adegbija, M.V., Fakomogbon, M.A. and Daramola F.O. (2012). The new technologies and the conduct of e-examinations: A case study of National Open University of Nigeria. *British Journal of Science Jan. 2012, 3 (1) 59-66.*

2. American Association for the Advancement of Science (AAAS) (2013). Science and diplomacy. AAAS Centre for Science Diplomacy.

3. Ekpenyong, L.E.(2010). Challenge of Assessment in Vocational and Technical Education: *Journal of Educational Assessment in Africa 28, 361.*

4. Federal Ministry of Education (2011). The State of Education in Nigeria Beyond Access.

5. Federal Republic of Nigeria (2004), National Policy on Education, 4th Edition 2004.

6. Gamble, J. (2004): A knowledge perspective on the Vocational Curriculum.

7. Hornby A.S. (1995). *Oxford Advanced Learner's Dictionary of Current English.* Oxford: Oxford University Press. 5th Ed.

8. Ikuenobe, A.F. (2014). Comparative analysis of resources in technical colleges under different proprietorship in southern Nigeria. Unpublished Thesis submitted to the Department of Educational Administration and Foundation, Faculty of Education, University of Benin.

9. International Baccalaureate Organisation (2005). Guide to programme evaluation. Peterson House UK *Retrieved from www.ibo.org on 1st August, 2012.*

10. Juma, C. (2013). Forging new diplomacy bonds through science and technology. Technology policy innovation at work.

11. Kwakpovwe, C.E. (2014). Our Daily Manner: The very urgent message. Lagos: Manner Resource Centre, 14(4,5,6), 40.

12. Lynn Mytel2defr3vfr3gtrgt3gtgtvfvfka (2002). Creating opportunity for learning and innovation through trade and transfer of technology. *General 20 April.*

13. Miliband (2010). Urges greater for science diplomacy. British Royal Society.

14. Nworgu, B.G. (2010). The challenges of Quality of Assessment in a Changing Global Economy, *Journal of Educational Assessment in Africa 28, 17-18.*

15. Okafor (2011). The role of votional and technical edution in manpower development and job creation in Nigeria. *Journal of Research and Development, 2(1), 156.*

16. Science Diplomacy. (Wikipedia, 2014). Modified on 12 March, 2014.

17. The Royal Society and American Association for the Advancement of Science (AAAS) (2010). New frontiers in science diplomacy: Navigating the changing balance of power. London: Techset Composition Limited.

18. UNCTAD (2003). Science and technology diplomacy: Concepts and elements of a work programme. Geneva: UNCTAD/ITE/TEB/Mics.

19. UNESCO (1999), Technical and Vocational Education and Training. A vision for the Twenty First Century – Recommendation. Paris

20. Gly, U. (2014) What is Science? Retrieved from http://www.gly.uga.edu/railsback/1122science2.html

21. William, E. (2013). Role of science in the third millennium. 46th Session of the Erice international seminars, Siclly.

*Chapter 9*

# Nigeria's Technical Aid Corps Scheme: A Model for Science and Technology Diplomacy in Developing Countries

*Bolarinwa Olugbemi*

*Deputy Director,*
*Policy Analysis and Development,*
*Raw Materials Research and Development Council,*
*Abuja, Nigeria*
*e-mail: drolugbemi@yahoo.com*

## ABSTRACT

Science and Technology Diplomacy could be described as the use of scientific and technological collaborations among nations to address common problems and to build constructive international partnership, through the deployment of scientific knowledge, products and experts. As part of her foreign policy, the Nigerian Government established the Technical Aid Corps (TAC) Scheme in 1987, as an alternative to direct financial aid for African, Caribbean and Pacific (ACP) countries. It was designed not only to provide manpower assistance in all fields of human endeavour but also to represent a practical demonstration of South-South cooperation. The objectives of the scheme include sharing Nigeria's know-how and expertise with other ACP countries, providing assistance on the basis of assessed and perceived needs of recipient countries, ensuring a streamlined programme of assistance to other developing countries, acting as a channel through which South-South collaboration is enhanced, establishing a presence in countries which, for economic reasons Nigeria has no resident diplomatic mission, promoting cooperation and understanding between Nigeria and recipient countries, facilitating meaningful contact between youths of Nigeria and those of the recipient countries, and complementing other forms of assistance to ACP countries. This programme cuts across diverse fields, with volunteers selected on the basis of qualification, expertise, competence and the needs of recipient countries. Adopting this model as a platform for science and technology diplomacy in NAM member states will further enhance development, cooperation, collaboration, exchange of knowledge and technology and such other benefits. The Raw Materials Research and Development Council, an agency under the supervision of the Federal Ministry of Science and Technology, with its expertise in developing Nigeria's raw materials resources has produced research products and technological innovations that could be deployed to other member states along the model of the Technical Aid Corps Scheme. This effort will assist in addressing issues contained in the Millennium Development Goals (MDGs), such as education, health, poverty alleviation, biodiversity, environment sustainability, climate change, security, global partnership for development, etc., in these countries. Authorities and Heads of Government of member states should adopt and endorse this scheme, through a pragmatic policy instrument that will ensure effective implementation of this concept. The NAM S&T Centre shall coordinate this programme.

*Keywords: Diplomacy, Science, Technology, Innovation, Technical aid corps scheme, Nigeria.*

# 1. Introduction

Since the end of the Second World War, the major powers had devised and perfected some core diplomatic weapons to influence and direct the course of world affairs. These instruments include economic and political diplomacy, membership and leadership of statutory influential United Nation's bodies and commissions as well as the power of Veto at the Security Council.

In addition, these super powers had also devised a pragmatic programme of deploying their knowledge, scientific and technological products, resources, culture, etc. to developing countries. These deployments are not without their prices and sovereign implications on the long run.

In order to be part of the global activities as well as provide the platform for economic, social, political, scientific and technological development, developing countries, under the aegis of the Non Align Movement came together to chart a course for their own destiny.

In view of the growing needs to develop other influential instrument of diplomacy aside the traditional ones, which are becoming less effective these days, world attention is being focused on science and technology as a diplomatic tool to further improve on bilateral and multilateral cooperation between and amongst nations.

Science and Technology Diplomacy can be described as the use of scientific and technological collaborations among nations to address common problems and to build constructive international partnership.

According to UNCTAD (2003), the term Science and Technology Diplomacy is used to mean the provision of science and technology advice to multilateral negotiations and the implementation of such results at the national level. It is now clear that advances in science and technology are becoming key drivers in international relations and negotiations.

International diplomacy now recognizes that scientific knowledge is becoming increasingly specialized and requires expert inputs into international negotiations, while at the same time requires the ability to integrate the divergent disciplines that are needed to solve specific problems.

The importance of science diplomacy was aptly captured by the former British Prime Minister, Gordon Brown at a two day meeting on "New Frontiers in Science Diplomacy" organized by the Royal Society in partnership with the American Association for the Advancement of Science (AAAS) on 1-2 June 2009, where he called for a new role for science in international policy making and diplomacy in order to place science at the heart of the progressive international agenda.

Since independence in 1960, Nigeria has adopted an Afro-centric foreign policy, with aid and technical assistance at the centre of her external relations with other countries, particularly within the African continent. The Federal Government of Nigeria in 1987 established the Technical Aid Corps (TAC) Scheme as a foreign policy tool that should serve specific national interest. The scheme involves the deployment of technical experts from Nigeria to needy African, Caribbean and Pacific (ACP) countries, to assists in the development of the recipient countries under mutually agreed terms. These experts include medical doctors, nurses, scientists, engineers, technologists, lecturers, architects, legal practitioners and other professionals.

This scheme is one of the best examples of a successful diplomatic instrument, which member countries of the Non Align Movement need to adopt and adapt in order to foster bilateral and multi-lateral relationships, spill-over knowledge and unlock the potentials in these countries through skills exchange. It is currently the only viable volunteer service operated by an African country.

## Overview of Nigeria's Technical Aid Corps (TAC) Scheme

The TAC scheme is a technical cooperation platform between Nigeria and African, Caribbean and Pacific (ACP) nations. It is an alternative to direct financial aid, designed for sharing know-how and expertise with other ACP countries. Nigeria engages the services of her nationals from various sectors such as Medicine and the Academia to execute this programme in the recipient countries.

The Programme acts as a channel through which South-South collaboration is enhanced, through streamlined programme of assistance to other developing countries. It shows enormous amount of local ownership and knowledge transfer from Nigerian experts to participating personnel of the recipient countries.

TAC has played a cardinal role in cementing relations between Nigeria and the beneficial countries, and, on a wider scale, creates an atmosphere of partnership where it will otherwise not exist.

## 2. Background

The Federal Government of Nigeria established the Technical Aid Corps (TAC) Scheme in 1987, as an alternative to direct financial aid for African, Caribbean and Pacific (ACP) countries. It was designed not only to provide manpower assistance in all fields of human endeavour but also to represent a practical demonstration of South-South cooperation.

## Key Objectives of the Scheme

   i. Sharing Nigeria's know-how and expertise with other ACP countries

   ii. Giving assistance on the basis of assessed and perceived needs of recipient countries

   iii. Ensuring a streamlined programme of assistance to other developing countries

   iv. Acting as a channel through which South-South collaboration is enhanced

   v. Establishing a presence in countries which, for economic reasons Nigeria has no resident diplomatic mission

   vi. Promoting cooperation and understanding between Nigeria and recipient countries

   vii. Facilitating meaningful contact between youths of Nigeria and those of the recipient countries

   viii. Complementing other forms of assistance to ACP countries

## Implementation of the Scheme

This scheme is administered by the Directorate of Technical Aid Corps within Nigeria's Ministry of Foreign Affairs. The programme calls for the provision of Nigerian expertise to needy member states under the Commonwealth Assistance Programme (CAP) and managed by the Directorate of TAC in Nigeria.

Volunteers are chosen from a pool of Nigerian citizens that applied to the Directorate of TAC, and these applicants are carefully selected for their skills, competence and knowledge in a particular field. Selected volunteers stayed for upwards of 2-4 years in the recipient countries and are completely funded through Nigerian public funds.

## Notable Outcomes/Achievement

   i. The scheme has received commendations from recipient countries and other members of the international community

   ii. Sierra Leone has been a beneficiary of TAC for the past 21 years and has received not less than 150 volunteers from Nigeria

   iii. TAC has played a significant role in cementing existing relations between Nigeria and beneficiary countries and creates partnership where it has not existed

   iv. Between 1987 when the scheme took off and now, over 4,000 TAC volunteers had been deployed to 33 countries, thereby providing a clear demonstration of Nigeria's foreign aid and technical assistance direction and policy

   v. The scheme has attracted attention from the Commonwealth Secretariat, culminating in the signing of a Memorandum of Understanding (MoU) with the Nigerian Government in 2003. Other international organizations that had shown interest in the scheme include the United Nations (UN) Volunteer Service and the Japanese Agency for International Cooperation (JAIC).

vi. In 2002, the Nigerian Government agreed to send medical volunteers to the Jamaican Government upon request from that country for a period of 2 years

vii. The Caribbean Island of Belize received 29 medical professionals under this scheme in March 2006. These professionals include doctors, nurses, pharmacists and laboratory technicians.

viii. In November 2006, 110 medical personnel, comprising nurses, doctors, pharmacists, physiotherapists and radiographic technicians were sent to Jamaica to assist in their heath sector.

ix. The Republic of Uganda also received 40 Nigerian medical personnel and deployed them to the Kampala International University Teaching Hospital as workers and teachers.

x. In 2009, the Nigerian Government released the sum of $70 million for the deployment of 132 highly qualified professionals to Sierra Leone, following the request made by the country's President, Ernest Koroma.

## 3. Policy Framework for the Adoption of the TAC Model for S&T Diplomacy in NAM Member States

i. At the high level of Head of Government of member states of NAM, a policy instrument should be conceived and adopted to facilitate exchange of skilled personnel in S&T within member countries along the framework of Nigeria's Technical Aid Corps Scheme.

ii. A data base of personnel in S&T in member countries of NAM should be provided and stored in larger retrieval platform at the NAM S&T Secretariat for ease of engagement.

iii. Requests from member states should be routed through the NAM Secretariat to the volunteering countries for proper monitoring and management.

iv. Rules of engagement should be formulated, agreed upon and endorsed by Heads of Government of member states

v. Each volunteering nation shall be responsible for full funding of its personnel, while recipient countries will provide accommodation, medical expenses/insurance and local transportation to these volunteers.

vi. All member countries shall participate in this programme

vii. Duration of programme per batch shall be for a period of 2 years in the first instance and a further 2 years, based on request from the recipient country.

viii. The exchange programme should be restricted to scientific and technological expertise including lecturers in these fields.

    ix. Artisans and technicians in the above fields shall also be encouraged to participate in this programme.

    x. The scheme may be tagged "NAM S&T Aid Corps Scheme or NAM S&T Volunteer Vanguard, or any such name or acronym that shall be agreed upon by the Authorities and Heads of Government of member states of NAM.

    xi. Member countries of NAM should adequately fund the scheme to enable it run smoothly

    xii. Research institutions and knowledge-based centres in the field of science, technology and innovation in member states should be the primary source of these volunteers.

    xiii. Embassies and High Commissions in member states should be mandated to maintain an S&T desk, with the incumbent being an influential international figure in the S&T family.

## 4. Raw Materials Research and Development Council's R&D Initiatives and Tools for Diplomacy

### National Research and Development Programme

The Raw Materials Research and Development Council (RMRDC), was established principally to promote and fast-track industrial development and self-reliance through the maximum utilization of local raw materials. The mandates of the Council as enshrined in the enabling Decree (now Act) No. 39 of 1987 are:

    ☆ To draw up policy guidelines and action plans on raw materials acquisition, exploitation and development.

    ☆ To review from time to time raw materials resources availability and utilization, with a view to advising the Federal Government on the strategic implication of depletion, conservation or stock-piling of such resources.

    ☆ To encourage the growth of implant research and development capabilities

    ☆ To organize workshops, symposia and seminars designed to enlighten the public on raw materials development and solutions discovered from time to time

    ☆ To advise on adaptation of machinery and process for raw materials utilization.

    ☆ To provide special research grants for specific objectives and device awards or systems for industries to achieve breakthrough or make innovation and inventions.

    ☆ To encourage the publicity of research findings and other information relevant to local sourcing of raw materials.

In consonance with its mandate, RMRDC instituted a National Research and Development Programme in 1990 to promote the development of local raw materials and industrial technology, especially indigenous technology, by supporting major R&D activities in research institute, universities, polytechnics, private and public R&D outfits. Through this, the Council brings the resources of the scientific and technical community in public and private sector laboratories to bear on the national problem of development and utilization of local raw materials.

The Programme was aimed at identifying and pursuing applied research and development opportunities that offer the greatest potentials for commercialization by industry, with emphasis on increased local content and value-added manufacturing.

The specific objectives of the programme include:

☆ Acquisition and adaptation of technologies

☆ Exploitation of local resources to replace imported raw materials

☆ Upgrading of indigenous technologies to facilitate efficient and rapid production of goods and services, based on local resources.

☆ Creation of new processes, products and technologies.

## Research Funding

In its two and a half decades of existence, the Council has funded about 114 R&D projects addressing the needs of various sectors of the manufacturing industry in Nigeria. It is important to stress that the nature of R&D enterprise is such that some investigations may fail to yield desired results, some may open new areas for further investigation and some may yield the expected results. The following are some of the successful RMRDC sponsored R&D Projects:

### Design and Fabrication of Spray Dryer for Small Scale Industry

This research led to the design and fabrication of a special dryer for use by small scale food industries. The equipment is suitable for drying milk, food flavour and spices, etc. The applicability of this equipment cuts across several industries and if fully utilized, would save the nation huge foreign exchange.

### Investigation of Medicinal Plants Used as Traditional Medicines for Diabetes and Snake Bites

This research produced active ingredients extracted from selected local plants found to be effective against diabetes and snake poisons. This result, when fully commercialized would reduce Nigeria's dependence on imported anti-diabetics and snake venoms.

### Developing Kapok (Auduganrinmi – Hausa) as Textile Fibre

This research sought to determine the spin ability and wearability of kapok fibre in blend with other fibres for use as textile fibre. However, the product of the research was found to be useful in the manufacture of carpet underlay. When this is commercialized, there would be a reduction in imported raw materials by the textile sector.

### Production of Soy Sauce

Aspergillus was used to ferment a mixture of soya – bean and sweet potato extracts to produce food flavour. The flavour was found to be very good, consumer friendly and capable of reducing importation of food flavours, if adopted by local food industries.

### Development of Long Staple Cotton for the Nigerian Textile Industry

The project resulted in the development of four varieties of long staple cotton, three of which were distributed to farmers under the Council's boosting programmes. Nigeria has now joined the league of long staple cotton producing nations. Thus, Nigerian Cotton exporters stand to reap more from exporting the commodity. Statistics of foreign exchange earned from the export of this commodity and its derivative products is put at N6.8 billion for 2005 and N4.5 billion for 2006.

### Production of Furfural Urea Fertilizer from Maize Cobs

Furfural produced from maize cob residue and mixed urea to produce fertilizer. Its application was found to have increased agricultural yield and consequently boost production of agro industrial fertilizer and save valuable foreign exchange for Nigeria.

### Utilization of Trona Residue as a Bleaching Earth

This research, which was designed to find local substitute to bleaching agent currently wholly imported, is a critical raw material in the chemical, pharmaceutical and food industries. The bleaching earth, produced from trona residue, successfully bleached oils from groundnut, shea butter, palm kernel, and red palm oil. The National Research Institute for Chemical Technology, Zaria, collaborated with RMRDC on this project and with some companies along the Zaria – Kano axis of Nigeria. Full commercialization of the result would save the nation a lot of foreign exchange.

### Bioactivity Guided studies of Acalypha Species

Extraction and purification of the active principles from this plant was achieved and the compound found to have anti-fungal and anti-bacterial properties. The result of this research has led to increased local production of anti-bacterial and anti-fungal products in the Nigerian market.

### Preparation and Characterization of Alkyd Resins Using Rubber Seed Oil

Alkyd resin produced from rubber seed oil was subjected to industrial test and confirmed to be useful for paint and adhesive manufacturing industries. If fully developed, this will save the Country a lot of Foreign Exchange being expended annually on importation of alkyd resin.

### Design, Construction and Testing of a Motorized Briquetting Machine for Wood and Agricultural Wastes

This research was designed to convert waste to wealth and provide an alternative source of energy. The equipment produced was used to briquette saw dust successfully and today, the briquette industry is gaining ground in Nigeria. There are firms producing briquette from other agricultural wastes including rice husks.

## On – Farm Trials of Alternative Formulation of Livestock Feeds

Some alternative livestock feeds were successfully tested in selected farms in Kwara State. The significance of this research is its potential to classically reduce the cost of livestock feeds in Nigeria as against the cost of imported feeds.

## Production and Commercialization of Dried Mango Chips Using a Locally Fabricated Multipurpose Dryer

An organized multipurpose dryer was fabricated and used to dry mango chips

Other Successful R&D Products by the Council include the following:

☆ Extraction and Evaluation of calabash Seed oils and the use of the Lipid Free Extract as animal feeds

☆ Production of Cellulose from Agricultural wastes.

☆ Development of Vegetable Curried Blood Meal as Raw Material for the Feed Mill Industry.

☆ Production of Pharmaceutical Grade Talc from Locally Sourced Talc Deposits.

☆ Development of Calcined Kaolin as a Partial Substitute for Titanium Dioxide in Paint Making Industry.

☆ Development of Wood Seasoning Kiln for the Furniture Industry.

☆ Conversion of Palm Kernel Wastes into Mineralized Organic Fertilizer.

☆ Production of Potash from Cocoa Husks.

## Pilot Plants

Following the successes recorded in some of the R&D projects sponsored by the Council, it became necessary to demonstrate their technological feasibility and economic viability on an industrial scale. To achieve this, some of the projects were upgraded to pilot scale. With this approach, it would be possible to establish small factories which run on a semi-commercial scale, utilizing local raw materials.

The Council made considerable progress in the establishment of the following pilot plants in various parts of the country:

☆ Full- fat Soya beans Processing

☆ Fish Smoking Equipment

☆ Glazier Putty Plant

☆ Sorghum Malting Plant

☆ Castor Oil Plant

☆ Organo-Mineral Fertilizer Plant

☆ Beniseed Oil Processing Plant

☆ Coal Carbonization/Briquetting Plant

☆ Essential Oil Plant

☆ Kilishi Processing Plant

☆ Shea Butter Production Plant

☆ Salt Processing Plant

☆ Proto-type Automated Clearing Loom

☆ Wood Seasoning Kiln

☆ Flash Dryer

☆ Moringa Water Treatment Plant.

Another initiative of the Council, aimed at building indigenous engineering capacity, is promoting the application of design parameters in equipment fabrication. Most equipment fabricators in Nigeria lacks design parameters, a situation which has compromised their efficiency and precision in reproducibility. Consequently, the Council instituted an annual National Design Competition on Process Equipment. To sustain this design competition programme, RMRDC has Promoted Public-Private Collaboration, which has resulted in the setting up of a National Foundation for Process Equipment Development (NFPED).

## 5. Tool for Diplomacy

Diplomatic engagements by nations has gone beyond the frontiers of national interests, sovereignty, military supremacy, economic sanctions, nuclear proliferation treaties, climate change and global warming, pollution, wars, hunger, natural disasters, etc., to knowledge spillovers, telecommunication technologies, high speed and highly complex technological prowess, knowledge sharing, product and cultural exports, etc.

The developed countries have been deploying their technologies since the end of the Second World War, to the developing and economically weak nations. These efforts have to a large extent, assisted these countries improve on their living conditions, educational facilities, human and natural resources, as well as their economies, politics, governance and development. Most of these activities come by way of aids to the developing countries.

Using science and technology as a tool of diplomacy is rather a new concept in Africa and to a reasonable extent, in the developing world.

In Nigeria, a lot of scientific research, leading to innovations and inventions had been carried out. These excellent works however reside mostly in the Universities and research institutes and are not acknowledged by the political class.

There are about 129 Universities in Nigeria (NUC), an almost equal number of Polytechnics and Colleges of Education and about 74 Research Institutes covering various fields of Science, Agriculture, Engineering and Technology, Food, Health, Chemical, Pharmaceuticals, etc. These institutions engage the best hands in the field and are turning out innovative research results and products.

Exporting these knowledge and products globally should now engage the attention of the political class, the bureaucrats and autocrats to begin to utilize scientific and technological breakthroughs as weapons of diplomatic engagement with other nations of the world. The era of economic sanctions are numbered. We should turn to Science and Technology to influence the world.

## 6. Conclusion

The global world is moving towards a knowledge-driven economy and the place of science, technology and innovation is central to driving this process. Member states of the Non-Align Movement must not be left behind in this race for survival. The new wave of diplomatic engagement is shifting towards science, technology and politics. As observed by David Milband, the UK former foreign Secretary during the 2010 Inter Academy Panel of the British Royal Society, "the scientific world is fast becoming inter-disciplinary but the biggest interdisciplinary leap needed is to connect the world of science and politics" This is the new approach to world politics and international relations.

Members of the Non Align Movement could adopt the Nigerian Model of the Technical Aid Corps (TAC) Scheme that had been successfully deployed by Nigeria to foster greater relationship between her and the ACP countries since 1987. Products of scientific discoveries, technological innovations and personnel could be deployed within member states through a programme of exchange following the TAC model.

The Raw Materials Research and Development Council could be the springboard to deploy products of scientific and technological innovation from Nigeria to other member states of the Non Align Movement under a protocol of agreement endorsed by Authority and Heads of Governments of participating states.

## REFERENCES

1. Brown, G (2009). Romanes Lecture in Oxford. Available online at www. number10.gov.uk/page 18472

2. Juma, C (2013). Forging New Diplomatic Bonds Through Science and Technology. Belfer Center for Science and International Affairs.

3. Newsome, S.S (2010). A case Study in Science and Technology Diplomacy: Understanding Diplomats' Technical Competency and Interaction with Technical Experts. Massachusetts Institute of Technology.

4. NUC. List of Nigerian Universities. www.nuc.edu.ng

5. UNCTAD (2003). Science and Technology Diplomacy Concepts and Elements of a Work Programme. Pp. 26.

# — *Section III* —
# International Organisations and Networking

*Chapter 10*

# Enhancing National Capacities for Sustainable Development: The Case of Oceans, Seas and Developing Countries

*Venugopalan Ittekkot*

**Professor and Former Director,
Leibniz Centre for Tropical Marine Ecology,
University of Bremen, Bremen, Germany**
*e-mail: ittekkot@uni-bremen.de*

## ABSTRACT

Climate and other global changes are threats to the health of oceans and seas. These affect their potential to provide goods and services. Both threats and responses of such systems are trans-boundary in nature. Measures to conserve and sustainably use oceans and seas have to based on a better understanding of their dynamics and appropriately structured policies at the national, regional and international levels. Many developing countries lack the needed capacity to assess, develop and implement such integrated measures. Experts in such countries should take note uncertainties identified by science while dealing with trans-boundary challenges. These in turn could determine consensus in policy making to deal with such challenges mutually reinforcing action at the local, regional and global levels.

*Keywords: Oceans and seas, Capacity development, Developing countries, Sustainable development.*

## 1. Introduction

The outcome document of Rio+20, "the Future We Want" recognizes "the critical role of Oceans and Seas to sustain Earth's life-support systems" and that "careless exploitation of the oceans and their resources risks their ability to provide food, other economic benefits and environmental services" and "stresses the importance of conservation, sustainable management and equitable sharing of marine and ocean resources".

Goods and services provided by oceans and seas are considered to have a GDP potential of about a trillion dollars. About 500 million jobs come from ocean sectors such as for example, fisheries and aquaculture, shipping, energy and tourism (UNDP 2012a, b). This economic potential is under threat from increase in atmospheric $CO_2$ and projected global warming (IPCC 2007). These global changes are frustrating the efforts of many developing nations that wish to promote use of oceans and seas for national development initiatives through the "blue economy" framework.

NAM member countries have jurisdiction over approximately 60 per cent of the world's coastline and their associated marine areas, some of which are hotpots of global marine biodiversity. Their enormous economic potential remains largely untapped or underused for national development due to inadequate or lack of needed capacity. This paper addresses some aspects of national capacities in developing countries and discusses possible measures to enhance them. Included is information from a recent survey among a selected group of developing countries including several NAM S&T Centre member countries (see Ittekkot 2013, 2014).

## 2. National Capacities and Needs

Capability to make the best use of the Oceans and Seas varies among developing countries. Some have created a base for the study and use of their oceans and seas by investing in human resources and infrastructure. Most however, lack the means or resources to adequately respond to the scientific, social, economic and political challenges related to conservation and sustainable use of oceans. There are deficits in such areas as education and research on oceanography (scientific and technical), infrastructure, marine policy making; skills to comply with international Conventions and Treaties and to benefit from related regional and international engagement.

Inadequate resources constrain capacity development actions in many countries. Very often there is no national coordinating agency for the use of oceans. Related responsibilities are divided among many ministries and departments. This fragmentation results in conflicting sector-oriented approach that does not serve an integrated national priority. This in turn, affects resource allocation and investments in much needed human capacity building and infrastructure. Capacity development efforts have therefore remained sporadic with overdependence on project-oriented, short-term external support. This tends to only meet the needs of donors needs and mostly without opportunities for sustained long term development at national level.

In this context, the establishment of the new Coordinating Ministry for Maritime Affairs by the newly-elected Indonesian Government is a potentially beneficial change. The Ministry will coordinate marine-related activities of other ministries responsible for sectors such as transportation, fisheries, tourism as well as energy and mineral resources with their potential to generate revenue (http://www.

thejakartapost.com/news/2014/11/10/building-foundation-maritime-ambition.
html).

## Education and Research

Expertise for oceans and seas in most developing countries resides in academic institutions. Very often however this is not adequately used or supported, Generally there is a worsening of conditions under which traditional institutions of higher education in many developing countries function. Diversion of public support for quick-fix mechanisms offered through private initiatives as well as the fiercely competing foreign educational institutions with their renewed strategies of internationalization has been critical. Existing Universities and Technical Institutions with a potential for excellence in trans-disciplinary education need to be identified and supported so that they can upgrade skills and infrastructure.

Developing countries could take advantage of experience of countries such as India. In the first two decades after British rule, India began systematically supporting indigenous Science and Technology Institutions, which it considered had the potential for excellence and for contributing to national needs and priorities. By an Act of Parliament in 1961, the Government of India declared them to be Institutions of National Importance. International cooperation and support from Northern countries – Germany, Soviet Union, United Kingdom, and USA were sought and leveraged for this. The promotion of Indian Institute of Technology across India is one of the results. Although a fraction of graduates leave the country for destinations in the North, graduates of these Institutes continue to contribute to India's development.

Retaining developed human capacity is another problem. In developing countries, government departments and academic institutions still remain the major employers of fresh graduates. But there is a high turnover of employees because they leave for other sectors, where jobs are more lucrative. To some extent international (global, regional and bilateral) collaborative programs help to retain newly-qualified personnel. Such opportunities have been few and adequate national efforts are needed to create newer opportunities.

There are provisions in some countries for hiring foreign experts, though for short periods with their limitations. Here, developing countries could also turn to the pool of expatriate ocean experts active in developed countries. There are no systematic efforts to take advantage of this pool of experts, though some contribute to efforts in developing countries of their origin as advisors, within exchange programs and in guest positions. Needed are investments in new programs that could probably be implemented with the support of developed countries.

### Technology and Infrastructure

Exploration and use of oceans and seas require an infrastructure, sometimes rather expensive, that needs to be operated and regularly maintained. The Rio+20 Outcome document mentioned at the beginning also highlights "the need for transfer of technology" based on the set of criteria and guidelines developed by the Intergovernmental Oceanographic Commission (IOC). This is to help transfer of marine technology for the implementation and effective use of Part XIV of United Nations Convention on the Law of the Sea (UNCLOS). In the IOC Guidelines, Marine Technology refers to instruments, equipment, vessels, processes and methodologies required to produce and use knowledge to improve the study and understanding of nature and resources of the ocean and coastal areas (IOC, 2005). To make the best use of the transferred technologies and infrastructure, developing countries need a critical mass of technical experts in countries. However, technical education in marine-related fields gets inadequate attention.

Capacity development in the field of Oceans is often hampered by non-availability of research vessels and the lack of ship-based training and education programs. Some progress has been achieved by making use of transit times between research areas (research legs) of foreign research vessels for capacity development purposes, acquainting young scientists and students with modern oceanographic instruments and in their operation. For example leading German oceanographers have been involved in the mentoring of students from Sub-Saharan African coastal states on board German research vessels such as the METEOR and MARIA S MERIAN for short periods (IfM, 2011). This is developing into a regular exercise and is considered beneficial by many developing countries.

**Policy and International Engagement**

Until recently, issues of oceans and seas rarely found a place in national policies and development plans of most developing countries. Where present, they were tucked away within the overall national S&T and development strategy. Recognizing the economic potential of oceans and seas, some countries are developing national marine strategies for their conservation and management. International agencies under UN are supporting these efforts through advice and measures aimed at enhancing national capacities in the field.

Most developing countries are signatories of international conventions and treaties related to oceans and Seas. Inadequate information on these systems within their extended maritime jurisdiction, and lack of capacity for compliance weaken their position in regional and international processes and negotiations. Their current nature of participation is perceived as passive. Information from even this engagement does not reach a larger national community interested in the ocean sector. As a result, the available potential is not fully taken advantage of. Countries need to seek a more active role in setting regional and international agenda on the use of the oceans. For this purpose, they need national experts sensitive to the trans-boundary nature and conflict potential, appreciate scientific uncertainties, and build consensus in policy making towards sustainability at national, regional and global levels, as skilled negotiators in international fora.

## 3. Support from International Organizations

Many organizations have been successfully promoting the enhancement of scientific and technical capacity for the study of the Oceans (*e.g.*, Partnerships for the Observation of Global Oceans - POGO - http://www.ocean-partners.org, Scientific Committee on Oceanic Research- SCOR -http://www.scor-int.org),

Intergovernmental Oceanographic Commission – UNESCO/IOC -http://ioc-unesco.org). In the past, these activities were often based on programs and projects developed outside the regions, where capacity is being built and the emphasis very often has been on the use of tools and methods for ocean observations and data handling and management (see also Morrison *et al.* (2013); TWAS (2004)).

The activities of the Centre for Science and Technology of the Non-Aligned and Other Developing Countries (http://www.namstct.org) provide opportunities for talented young scientists from developing countries to spend time at advanced marine research institutions in the North. The contacts so established very often lead to follow-up "North-South" partnership projects in which young scientists work together on scientific issues of national importance. These activities thus set the context for transfer of marine technology through targeted Workshops (*e.g.,* Coastal Ecosystems: Hazards, Management and Rehabilitation in Indonesia) and annual Fellowship Programs in partnership with Institutions in the North.

Many among NAM nations have the capability to plan and implement advanced oceanographic research. These nations could join other interested NAM coastal states to initiate new international cooperative programs that specifically address marine scientific themes that are of relevance to regions. Such programs provide an opportunity to share talents, expertise and infrastructure. Such international programs addressing regional oceanographic themes have in the past contributed to strengthening oceanography both in developed and developing countries. One such program, the International Indian Ocean Expedition (IIOE) will be celebrating its 50th anniversary during 2015-2018.

The IIOE was planned and implemented by SCOR and UNESCO-IOC. The planning included nations bordering the northern Indian Ocean which recognized the opportunities and benefits from this international program. The benefits included working with some of the best oceanographers in the world and in furthering the development of oceanography in the Indian Ocean region. At the core of the capacity building activities was a Regional Centre, where scientists of international repute helped to build local capacity. The activities of the Centre and the capacity built there also contributed to the establishment of India's National Institute of Oceanography in 1965. The ingredients of IIOE's success were: (i) a challenging scientific theme it addressed – oceanography in a monsoon-affected region; (ii) participation of a group of international scientists with a passion for oceans including champions and lobbyists for the project in developing countries, who saw international cooperation as an opportunity and an instrument of development; as well as (iii) long-term commitment from governments in support of building organizational and institutional infrastructure. The IIOE contributed to the development of oceanography also in developed countries by providing a new platform for furthering the research capabilities of many scientists and oceanographic institutions there.

## 4. Conclusions

Developing countries do not confer the priority oceans and seas deserve in management strategies; except perhaps in the area of fisheries. As suppliers of diverse goods and services of enormous economic potential, oceans and seas are critical in fulfilling many of the goals of the post-2015 development agenda discussed across the world.

Addressing the deficits mentioned in this paper will require actions at national, regional and international levels. In some cases, regional investments, networking and sharing of facilities and infrastructure could be beneficial. NAM S&T Centre could bring together member countries within an Oceanographic Network, where countries can support each other in the conservation and use of oceans and seas under their national jurisdiction. This could be in the field of education and research, national and regional policy making or in the design and implementation of regional oceanographic programs.

National investments, in education, infrastructure as well as designing a national strategy for the conservation and use oceans and seas however meagre will go a long way towards developing and sustaining capacities that are rooted in local needs, priorities and traditions. This will also help nations to better leverage international cooperation and partnerships for national development.

## 5. Acknowledgments

I would like to thank Prof. Arun Kulshreshtha for giving me the opportunity to participate in the Workshop on "Science and Technology Diplomacy in Developing Countries" organized by the NAM S&T Centre and for the Centre's support and cooperation in efforts towards capacity development in ocean research.

## REFERENCES

1.  IfM, Universität Hamburg/Leitstelle Deutsche Forschungsschiffe., 2011Research Vessel Maria S. Merian, Cruise Nr MSM 17 (23.9.11 – 22.12.2011) Training and Capacity Building (http://www.ldf.uni-hamburg.de/de/merian/ wochenberichte/wochenberichte-merian/msm19/msm19-expeditionsheft.pdf date of access)

2.  IOC Advisory Board of Experts of the Law of the Sea., 2005. IOC Criteria and Guidelines on the Transfer of Marine Technology (CGTMT), UNESCO Paris, France.

3.  IPCC 2007: Synthesis Report, Contribution of the Working groups I, II and III to the Fourth Assessment Report of the Intergovernmental Panel on Climate Change. Core Writing Team, Pachauri, R. K. and Reisinger, A. (Eds), IPCC, Geneva, Switzerland.

4.  Ittekkot V., 2013. Capacity Development Needs: Observations from an IOC Survey, Presentation at the Expert Group Meeting on Oceans, UN DESA, April 27, 2013, (http://sustainabledevelopment.un.org/content/ documents/1761ITTEKKOT_EGM_OCEANS_April27_2013.pdf) based on "Baseline Study for an Assessment of National Capacities and Needs in Marine Research Observation and Data/Information Management", (http://unesdoc. unesco.org/images/0022/002268/226864e.pdf date of access)

5.  Ittekkot, V., 2014. Oceans, Seas and Sustainable Development: Preparedness of Developing Countries. Environmental Development, http://dx.doi. org/10.1016/j.envdev.2014.12.001

6.  Morrison R.J. *et al.* (other authors....)., 2013. Developing Human Capital for Successful Implementation of International Marine Scientific Research Projects. Marine Pollution Bulletin, 77: 11–22.

7.  National Research Council., 2008. *Increasing Capacity for Stewardship of Oceans and Coasts: A Priority for the 21st Century.* National Research Council of the National Academies, The National Academies Press, Washington, D.C., USA

8.  TWAS., 2004. Building Scientific Capacity: A TWAS Perspective. Report of the Third World Academy of Sciences. Third World Academy of Sciences, Trieste, Italy

9.  UNDP., 2012a. Catalysing Ocean Finance Volume I Transforming Markets to Restore and Protect the Global Ocean. United Nations Development Programme, New York 10017, USA, 54pp

10. UNDP., 2012b. Catalysing Ocean FinanceVolume II Methodologies and Case Studies. United Nations Development Programme, New York 10017, USA, 76pp

11. Zahuranec, B., Ittekkot, V. and Montgormery, E., 2014. Preface In: Science and Technology Diplomacy in Developing Countries, Zahuranec, B., Ittekkot, V. and Montogomery, E. NAM S&T Centre, Daya Publishing House, New Delhi, pp. xiii-xv.

*Chapter 11*

# Science and Technology Diplomacy in the Area of Nanotechnology

*Radhika Tandon*

*Research Associate,*
*NAM S&T Centre, New Delhi, India*
*e-mail: rdhk.tndn@gmail.com*

## ABSTRACT

Availability of technology and consequent products is an indicator of prosperity of any nation. The world is fast changing and each day new technologies are appearing on the horizon that are transforming the 'improbable' to 'probable'; are helpful in combating diseases, poverty, hunger, scarcity of clean water and air, energy crisis, natural disasters, etc.; are sustainable and less expensive; and which excel the quality of the products hitherto available. However, most of this work is concentrated in the global North. The developed countries make extensive use of North-North cooperation in order to accelerate their technology growth, but to cut on cost these countries also look towards South to take advantage of the cheap labour, which often is also relatively highly industrious and generally working unperturbed by the national and local labour laws, to reduce the material costs, and in many cases, to bypass the intellectual property rights. Thus for converting the innovative ideas into useful technologies and accelerating the technology growth the developed countries need the cooperation with other developed countries. However, because of the availability of huge markets in the developing countries and for their own economic gains through commercial exploitation of these technologies, the developed countries adopt the North-South cooperation route. The developing countries on the other hand are relatively slow in their technology development and for them the bilateral, regional and multilateral cooperation either in South-South or North-South cooperation modes become crucial to benefit from the rapid technology advances.

Nanotechnology is the material transformation, innovation driven advancement and development of our times wherein the materials at nanoscale behave differently due to size variation and corresponding properties from the same material at the micro or macroscale. With the advent of achieving the desired materials properties by suitably tailoring and designing the nanomaterial, the nanoscience research has emerged as a monumental scientific endeavour worldwide. This has resulted in up-grading the sciences in the area of robotics, aeronautics and other transport technologies, high-performance computing, social media, software, cost-effective energy sources, nanobiotechnology including genetic engineering, etc. Moreover the present day world is facing unprecedented problems, which are expected to be resolved by finding new solutions through the nanotechnology route and offer efficient manufacturing with less resources and smaller waste and products with better functioning. Developed countries have been taking advantage of this knowledge since many decades, but the developing countries too have now slowly starting to become aware of the magic of nanotechnology. Globally, the total nanotechnology investment till date is estimated to be more than $20 billion. The patterns and mechanism for nanotechnology working programmes are differently adopted by individual countries. In this regard, the key economically potent countries like the USA, Japan, Germany and those in EU, China and Russia are heavily investing to support significant national priorities by taking initiatives in this aspect, for example, through agencies like NNI in the USA, Multiannual framework programmes in the EU, Nano S&T programmes of China and RusNano agency in Russia. The nanotechnology lead is further followed by groups such as EU-27, G-20, BRICS and ASEAN nations. But in the developing countries the R&D funding in nanotechnology in their research institutions, universities and even the private sector is rather scarce and abysmal, though they would like to build a pool of scientists and researchers with adequate expertise in this field and exploit various applications as may be required by them not only for the welfare of their people like in safe drinking water, energy, agriculture, medicine etc, but also in manufacturing areas, and enter into global competitiveness in the area of nanotechnology.

In a way, nanotechnology has now taken the shape of being a strategic area, but in which there is huge technology divide between the North and South. Technology transfer is required in nanotechnology, nano-devices and nano-engineering mainly from North to South. The North also is deeply interested in exploiting the markets available in the developing countries for their nano-products, which needs discussions and negotiations at various levels and consequently engaging each other through science and technology diplomacy. Further, because there is some level of mistrust in the developing countries about having been able to acquire 'real' technology from the developed countries even by paying high price, South-South cooperation for them becomes a key element in their survival strategy. All these cross-interests and desirability of promoting partnerships to fulfil their own agenda vis-á-vis other nations requires negotiation and engagement of the concerned parties including those who have negotiating skills, and at the same time, deep scientific understanding and knowledge. In summary, science and technology diplomacy in a 'super'-fast emerging area of nanotechnology is becoming of paramount significance to all countries in the world for their economic progress.

The present paper discusses the status of nanoscience and nanotechnology in many developed and developing countries as well as within the groupings of several countries, for example, BRICS, ASEAN and SAARC, and cooperation mechanisms adopted by them for the promotion of nanotechnology to meet their individual requirements.

*Keywords:* Nanotechnology, Diplomacy, Bilateral, Regional and multilateral cooperation, Innovation.

# 1. Introduction

Innovations and inventions have a critical impact in all the fields to achieve sustainable development. Although many new technologies are occupying the market research areas such as artificial intelligence of making intelligent machines; genetic engineering using biotechnology; but the most recent addition to our science and technology diplomacy toolkit is nanotechnology. The concepts of nanotechnology began about 30 years ago, when our tools to image and measure expanded to the nanoscale, observing that physicists, biologists, chemists, electrical engineers, optical engineers, and materials scientists were working on overlapping issues emerging at the nanoscale. Today this technology is not isolated with the developed world but has blended with the emerging nations and thus recognising the potential of nanotechnology, nations are making huge investments, initiatives and planned strategies to exploit the technology appropriately leading to bring tremendous benefits to its citizens. Also in order to address the global advances, the successful application of nanotechnology depends upon leveraging international collaborative research under bilateral and multilateral agreements with other countries. The features of nanotechnology extends to the study, design, creation, synthesis, manipulation, and application of functional materials, devices, and systems through control of matter at the nanometer scale- *i.e.*, at the atomic and molecular levels. Further, the materials at nanoscale behave differently due to size variation and corresponding properties from the same material at the micro or macroscale. The creativity of this technology relates not only to reduction in sizes but also changing or enhancing properties that could provide the potential for pursuing and developing both evolutionary as well as revolutionary applications.

In recent years, the advancements in science and technology (S&T) have no longer remained confined to the boundaries of conventional technologies; but the local and national governments across the world are putting efforts by engaging their researchers to explore the immense benefits of nanoscience and nanotechnology by initiating new programmes and schemes. Many countries, with investment from governments, as well as private companies and academic and scientific institutes are engaged in the research and development (R&D) in nanotechnology. Since the year 2000, when the U.S. National Nanotechnology Initiative (NNI) was announced, more than sixty countries have established their national nanotechnology programmes [Roco, 2004].

## Potential Capacity of Nanotechnology Effecting its Importance in S&T Diplomacy

Nanoscience is the study of phenomena and manipulation of materials at nano scale. A nanometer (nm) is one thousand millionth of a meter. [Drexler, 1986] At this scale the behaviour of a material differs in fundamental ways from that observed at the macro and the microscales. Nanotechnology is the design, characterisation, production and application of structures, devices and systems by controlling shape and size at this scale [The Royal Society and the Royal Academy of Engineering, 2004]. However, Nanotechnology is not new; it has been explored in the past and there have also been a few diverse applications. It is only the last decade that has seen significant increase in interest in the technology. This has been because of advances that have enabled development of tools that now allow atoms and molecules to be examined and probed with great precision. Improved fabrication and machining technologies that allow very high precision and accuracy are enabling converting of sciences into workable models and products with implications across a broad range of domains [Ogawa, 1982]. Two principal factors cause the properties of nanomaterials to differ significantly from other materials:

i. **Increased relative surface area**: As a particle decreases in size, a greater proportion of atoms are found at the surface compared to those inside. Thus nanoparticles have a much greater surface area per unit mass compared with larger particles, resulting in materials at these sizes being more reactive chemically than the same mass of material made up of larger particles.

ii. **Quantum effects**: Quantum effects can begin to dominate the properties of matter as size is reduced to the nanoscale. These can affect the optical, electrical and magnetic behaviour of materials, particularly as the structure or particle size approaches the smaller end of the nanoscale.

The major effects related to nanotechnology are improvements in information processing capabilities, development of novel engineered materials and improving functionality of materials. Integration of these at a future date would lead to more autonomous systems and nano-robotic applications. Elaborating, value chain of nanotechnology products could be divided into three major categories. These are nanomaterials, nano-intermediates and nano-enabled products. Nanomaterials are nanoparticles, nanotubes etc., *i.e.* the nano-scale materials in unprocessed form. Nano-intermediates are the intermediate products with nano-scale structures in them. Nano-enabled products are the finished goods with nanotechnology incorporated into them. As example, titanium dioxide nanoparticles with the photocatalytic activity are nanomaterials, but the paint or a coating prepared with them is a nano-intermediate. A car or a building with such a coating or paint will be the example of a nano-enabled product. As one can imagine the value increases many folds when moving from nanomaterials to nano-enabled products. Nanotechnology is commonly considered as promising in a wide-range of high-tech sectors, especially in the context of pressing global challenges such as those related to energy, health care, clean water and climate change. However, it is estimated that the total world population may rise to around 9-10 billion by 2050 with higher expectations and demand in increased consumption of water, food, energy and materials [UN-REDD, 2011]. For example, 1.1 Billion people are at the risk from lack of clean water. In connection to this, various collaborative linkages have been facilitated to eradicated the global problems, thereby a team of Indian and U.S scientist have developed carbon nanotubes filters that remove bacteria and viruses more effectively than conventional membrane filters [Rensselaer, 2004].

Consequently, the changing society will correspond to the need of radically new technologies. Further, it is believed that as the Nanotechnology governance will get institutionalised with increased globalisation and co-funding mechanism. It

**Table 11.1: The Different Generations and Frames of Nanotechnology Development.**

| | Four Generations | Generation Characteristics | Risk Governance Context |
|---|---|---|---|
| **FRAME 1** | **First Generation –** passive (steady function) nanostructures<br><br>e.g. nanostructured coatings and non-invasive; invasive diagnostics for rapid patient monitoring<br><br>*From 2000 -* | *Behaviour:* inert or reactive nanostructures which have stable behaviour and quasi-constant properties during their use.<br><br>*Potential risk:* e.g. nanoparticles in cosmetics or food with large scale production and high exposure rates. | *Current context for Frame 1 products and processes:* interested parties are seeking to develop knowledge about the properties of nanomaterials and their EHS implications so that risks can be characterised internationally. Debates are focused on the design and implementation of best practices and regulatory policies.<br><br>*Risk characterisation:* the nanoscale components of the nanoscale products and processes result in increased system component complexity.<br><br>*Strategies:* the establishment of an internationally reviewed body of evidence related to toxicological and ecotoxicological experiments, and simulation and monitoring of actual exposure.<br><br>*Potential conflict:* the question of how much precaution is necessary when producing the nanomaterials (focusing on changes to best practices and regulation) and over their use in potential applications. |
| **FRAME 2** | **Second Generation –** active (evolving function nanostructures)<br><br>e.g. reactive nanostructured materials and sensors; targeted cancer therapies<br><br>*From 2005 -* | *Behaviour:* the nanostructures' properties are designed to change during operation so behaviour is variable and potentially unstable. Successive changes in state may occur (either intended or as an unforeseen reaction to the external environment).<br><br>*Potential risk:* e.g. nanobiodevices in the human body; pesticides engineered to react to different conditions. | *Current context for Frame 2 products and processes:* interested parties are considering the social desirability of anticipated innovations. Debates are focused on the process and speed of technical modernisation, changes in the interface between humans, machines and products, and the ethical boundaries of intervention into the environment and living systems (such as possible changes in human development and the inability to predict transformations to the human environment).<br><br>*Risk characterisation:* the nanoscale components and nanosystems of the Frame 2 products and processes result in knowledge uncertainty and ambiguity. |
| | **Third Generation –** integrated nanosystems (systems of nanosystems)<br><br>e.g. artificial organs built from the nanoscale; evolutionary nanobiosystems<br><br>*From 2010 -* | *Behaviour:* passive and/or active nanostructures are integrated into systems using nanoscale synthesis and assembling techniques. Emerging behaviour may be observed because of the complexity of systems with many components and types of interactions. New applications will develop based on the convergence of nanotechnology, biotechnology, information technology and the cognitive sciences (NBIC).<br><br>*Potential risk:* e.g. modified viruses and bacteria; emerging behaviour of large nanoscale systems. | *Strategies:* stakeholders must achieve understanding, engage in discussion about ethical and social responsibility for individuals and affected institutions and build institutional capacity to address unexpected risks. Projected scenarios need to be explored that show plausible (or implausible) links between the convergence of technologies and the possible social, ethical, cultural and perception threats. A major challenge is that decisions need to be undertaken before most of the processes and products are known.<br><br>*Potential conflicts:* the primary concern of Frame 2 is that the societal implications of any unexpected (or expected but unprepared for) consequences and the inequitable distribution of benefits may create tensions if not properly addressed. These concerns about technological development may not be exclusively linked to nanotechnology but are, at least partially, associated with it and will impact upon stakeholder perceptions and concerns. |
| | **Fourth Generation –** heterogeneous molecular nanosystems<br><br>e.g. nanoscale genetic therapies; molecules designed to selfassemble<br><br>*From 2015/2020 -* | *Behaviour:* engineered nanosystems and architectures are created from individual molecules or supramolecular components each of which have a specific structure and are designed to play a particular role. Fundamentally new functions and processes begin to emerge with the behaviour of applications being based on that of biological systems.<br><br>*Potential risk:* e.g. changes in biosystems; intrusive information systems. | |

would bring tremendous benefits to developing nations if exploited appropriately. Mihail (Mike) Roco of the U.S. National Nanotechnology Initiative (NNI) has described four generations of nanotechnology development [Roco, 2001] and the world (especially developed/high income countries) is assumes as in Table 11.1. The current era, as Roco depicts it, is that of passive nanostructures, materials designed to perform one task. The second phase, which we are just entering, introduces active nanostructures for multitasking; for example, actuators, drug delivery devices, and sensors. The third generation is expected to begin emerging around 2010 and will feature nanosystems with thousands of interacting components. A few years after that, the first integrated nanosystems, functioning (according to Roco) much like a mammalian cell with hierarchical systems within systems, are expected to be developed.

Since the announcement of National Nanotechnology Initiative (NNI) by USA in 2000 almost every developed and developing economy has initiated national nanotechnology programs, wherein, promoting diplomacy. The world assumes nanotechnology development into four different generations; commencing from 2000 where the U.S National Nanotechnology Initiative (NNI) was announced, the initial investment was formulated in the world and since then rebasing the nanotech impact factor on the U.S (=100) gives a clearer picture of where we expect the technology to have the greatest impact [Cientifica Ltd, 2011]. Eventually, it is changing the foundations from macro to micro and now exploring the nanotechnology via collaborative networking that demands attention, knowledge, research and further commercialisation.

Broader societal implications have indicated that 2 million nanotechnology workers will be needed by 2015 worldwide, but National Science Foundation has estimated thus number to be 6 million by 2020, with 2 million jobs in the US alone. It is also estimated that $1 trillion products incorporating nanotechnology will be achieved by 2015. Figure 11.1 represents a plethora of opportunities and windows that globally exist [Roco, 2011].

In January 2010, the Royal Society and the American Association for the Advancement of Science (AAAS) [RS Policy document, 2010] noted that "science diplomacy" refers to three main types of activities:

a) **Science in diplomacy:** Science can provide advice to inform and support foreign policy objectives.

b) **Diplomacy for science:** Diplomacy can facilitate international scientific cooperation.

c) **Science for diplomacy:** Scientific cooperation can improve international relations.

Emphasizing, Diplomacy for science leverages development through International cooperation and overall, we can say that Diplomacy is an irreplaceable option and science is an irresistible force. In this relation, no country would like to be left behind in accruing the benefits of this emerging Nanoscience and Nanotechnology field, thereby; almost all the nations in the world are keeping pace with the S&T diplomacy in Nanotechnology.

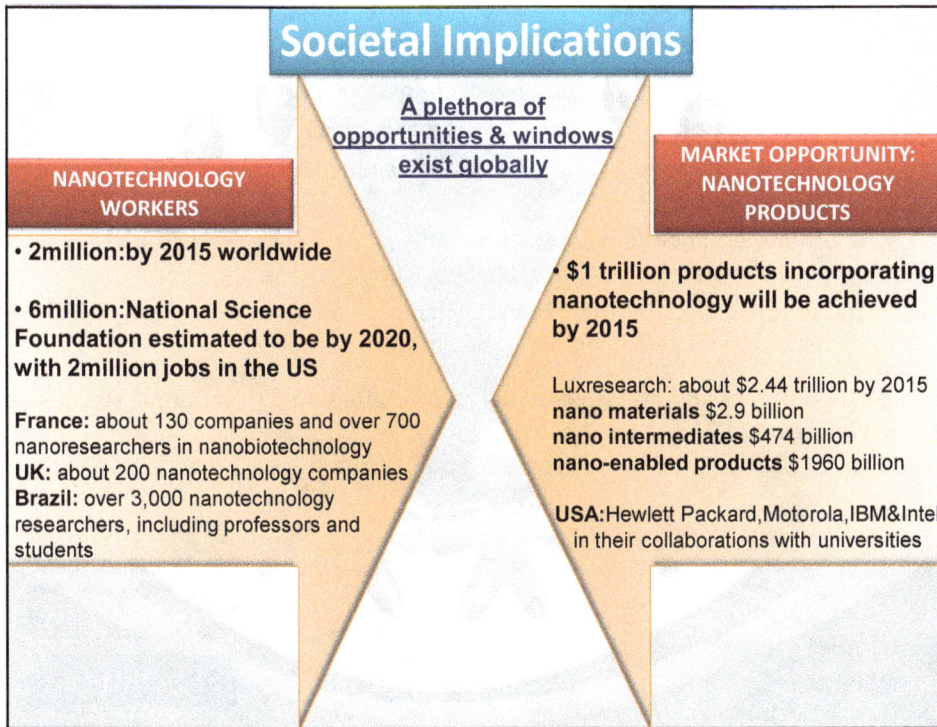

## Societal Implications

### A plethora of opportunities & windows exist globally

**NANOTECHNOLOGY WORKERS**

- 2million:by 2015 worldwide

- 6million:National Science Foundation estimated to be by 2020, with 2million jobs in the US

**France:** about 130 companies and over 700 nanoresearchers in nanobiotechnology
**UK:** about 200 nanotechnology companies
**Brazil:** over 3,000 nanotechnology researchers, including professors and students

**MARKET OPPORTUNITY: NANOTECHNOLOGY PRODUCTS**

- $1 trillion products incorporating nanotechnology will be achieved by 2015

Luxresearch: about $2.44 trillion by 2015
**nano materials** $2.9 billion
**nano intermediates** $474 billion
**nano-enabled products** $1960 billion

**USA:**Hewlett Packard,Motorola,IBM&Intel in their collaborations with universities

**Figure 11.1: Future Opportunities in Nanotechnology.**

## Collaborative Patterns, Proximity Influence and Global Initiatives

Advancements in nanotechnology are no longer confined to the individual nations but two styles exist in line with the international collaboration. Thus while USA, Germany, UK and Japan collaborate with a wide range of countries/region, Spain, Israel, Russia, Singapore and Taiwan are more selective in their collaboration partners [Zheng, 2013]. China's international collaboration partners were relatively narrow when compared with the USA and other western countries. During the past 20 years, as the time goes by, international partnerships density in nanotechnology is rising and increase in collaboration networking and cooperation among countries/regions has been prominently closer. Bi-lateral and multi-lateral partnerships and exchange programmes among various Asia-Pacific countries and Europe, US or other parts of the world are common place.

It is significant to mention about Global Nanotechnology Network (GNN). It is a diverse group of nano stakeholders worldwide with partner networks in Africa, Asia, Europe and the Americas. Network focus areas include: basic nano research, nanotechnology development, nano education and training, resource development, and development of a cyber infrastructure and database to serve the global nano community. Its mission is to serve as a platform for addressing shared global challenges through nanoscale science, engineering, technology development, and education [NCLT, 2009]. The goals of GNN are as follows:

★ Engage diverse partners across regions, disciplines and economic sectors

★ Create opportunities for collaborative research and education

★ Develop the next generation of global leaders in nanotechnology research and education

★ Promote nanoeducation as a means of increasing science literacy and public awareness of nanotechnology worldwide

★ Serve as a global clearinghouse in nano research and education, providing resources for the global nano community.

GNN activities and initiatives focus on four integrated programme strands as shown in Figure 11.2:

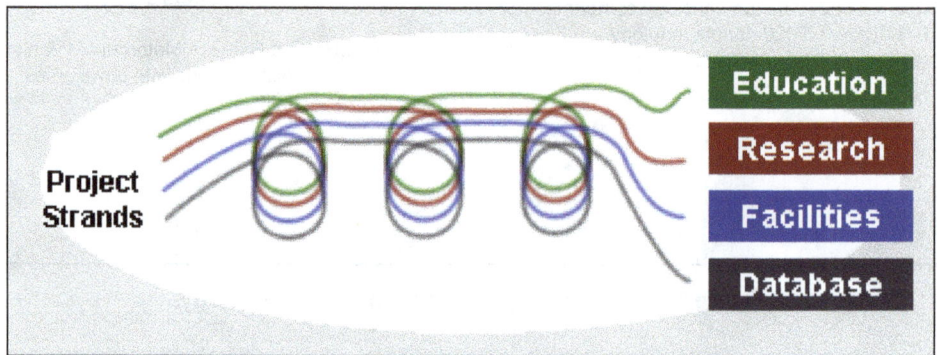

**Figure 11.2: Integrated Programme Strands of GNN.**

★ *Education:* Nanoeducation has excellent potential for raising science literacy. Nanoscale phenomena and cutting-edge nanotechnologies help make basic science concepts more interesting and relevant to students, and the inherent interdisciplinarity of nanoscience makes it an ideal focus for improving skills in Science, Technology, Engineering, and Math (STEM). GNN holds nanoeducation events that bring nano researchers and educators together to improve science education. GNN also helped launch the 'Global School for Advanced Studies', a global leadership development programme for young researchers around the world.

★ *Research:* Nanoscale science and technology can produce solutions to some of the world's most urgent problems - from clean renewable energy sources to technologies for cleaning our air, providing safe drinking water and curing diseases. GNN is a platform enabling researchers with complementary research capabilities to collaborate towards solving these pressing problems.

☆ *Facilities:* Nano research requires specialised equipment and close cooperation across fields and disciplines. GNN fosters sharing of nano-related facilities worldwide.

☆ *Database*: GNN is leading the efforts to develop a global database of nano research, facilities and education.

Figure 11.3 [Web: US NNI] [Food and Chemical Safety Committee, 2009] highlights other major grouping in international, regional and national arena.

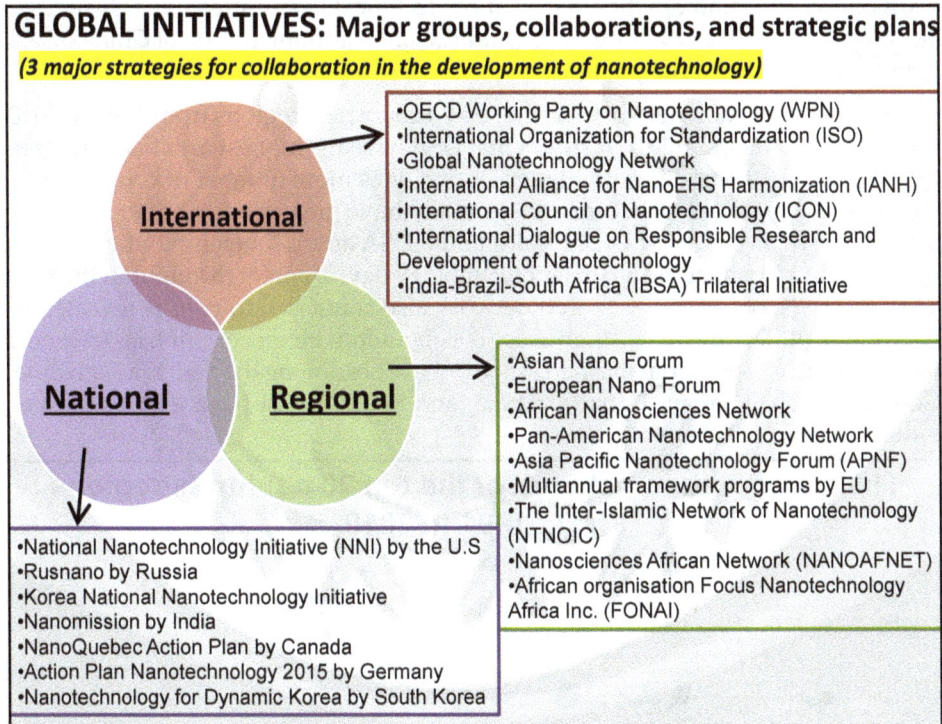

GLOBAL INITIATIVES: Major groups, collaborations, and strategic plans
(3 major strategies for collaboration in the development of nanotechnology)

International
•OECD Working Party on Nanotechnology (WPN)
•International Organization for Standardization (ISO)
•Global Nanotechnology Network
•International Alliance for NanoEHS Harmonization (IANH)
•International Council on Nanotechnology (ICON)
•International Dialogue on Responsible Research and Development of Nanotechnology
•India-Brazil-South Africa (IBSA) Trilateral Initiative

National    Regional
•Asian Nano Forum
•European Nano Forum
•African Nanosciences Network
•Pan-American Nanotechnology Network
•Asia Pacific Nanotechnology Forum (APNF)
•Multiannual framework programs by EU
•The Inter-Islamic Network of Nanotechnology (NTNOIC)
•Nanosciences African Network (NANOAFNET)
•African organisation Focus Nanotechnology Africa Inc. (FONAI)

•National Nanotechnology Initiative (NNI) by the U.S
•Rusnano by Russia
•Korea National Nanotechnology Initiative
•Nanomission by India
•NanoQuebec Action Plan by Canada
•Action Plan Nanotechnology 2015 by Germany
•Nanotechnology for Dynamic Korea by South Korea

Figure 11.3: International, Regional and National Global Initiatives in Nanotechnology.

Knowledge flow between nations strongly depends on factors such as: geographical proximity; being similar economic actors; functional closeness (regional, national and global level); belonging to same unit of administration; and reducing the transaction cost associated with the exchange of knowledge. Geographical proximity has impacted collaboration ties, for instance, US nanotechnology concentrates on the regional level and in general, the priority of collaboration in nanotechnology area is first within the continent. While considering technological proximity, the second preference for Asian countries is US and third one is Japan and China, two developed countries in Asia. Geographically, neighbouring Countries have many motivations to collaborate in developing science and technology. The clear examples are Azerbaijan, which chose Iran, the south neighbourhood of Azerbaijan for collaboration; New Zealand, which chose Australia to collaborate over other countries. Moreover, European countries are decentralized regarding the organisation and funding of nanotechnology research whereas US is in fierce competition with majority of technologies from synthetic biology to social networking. Russia alone is at a very high level stand by creating a joint-stock company Rusnano (formerly Russian Corporation of Nanotechnologies) owned by the government of Russia that aims at commercialising the developments in nanotechnology. China is an active participant in OECD working on manufactured nanomaterials, nanotechnology standardisation and characterisation involving the National Nanotechnology Standardization Technical Committee (NSTC) and the Technical Committee 279 under the standardization administration of China (SAC) [Cunningham and Werker, 2012].

The collaboration network for the top 20 countries/regions from 1991 to 2010 is represented in Figure 11.5. It shows that USA have the highest number of Degree Centrality (DC; the nodes with higher DC includes more number of connections), are more central to the structure and generally have greater potential to influence other nodes. The values of DC reveal that the USA and UK belong to the first tier of international collaboration in nanotechnology development throughout the past 20 years. The density of the overall network also continued to rise, as revealed in Figure 11.4, indicating increasingly close collaborations between countries/regions, which can be reflected from the increasingly dense ties among the nodes. Meanwhile, Germany, France, Japan, Canada, China, and South Korea have exerted greater influence steadily [Zheng, 2013].

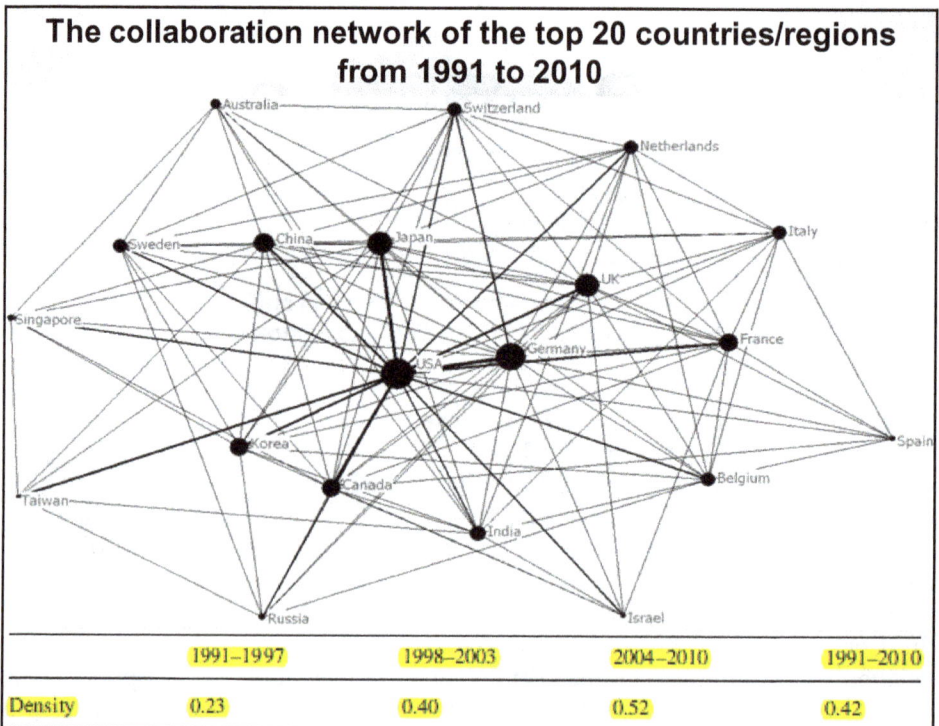

| | 1991–1997 | 1998–2003 | 2004–2010 | 1991–2010 |
|---|---|---|---|---|
| Density | 0.23 | 0.40 | 0.52 | 0.42 |

Figure 11.4: Collaboration Network for Top 20 Countries/Regions from 1991 to 2010.

Individually, measure of a nation's capabilities and resources for nanotechnology activity rely on indicators such as government funding, number of patents and publication and then the commercialisation prowess. It is observed that the US and Japan score the highest, directly followed by Germany with high nanotechnology activity and development strength; UK and France have high nanotechnology activity but low technology development strength; Switzerland and Sweden are involved in high general technology commercialisation; Netherlands and Italy have the least score [Forfas, 2010]. Nations have varying perspective in practical dealing on nanotechnology. Recalling the Clinton's dialogue, US aims at development of materials with ten times the strength of steel and shrinking all the information at the Library of Congress into a device the size of a sugar cube [Clinton, 2000]. The Rusnano's business strategy of 2008, states that the corporation is the "key coordinator of innovation policies designed to commercialize promising research and development projects in the field of nanotechnology" [Web: rusnano.com]. In India the great visionaries like Dr. Abdul Kalam and eminent scientists like Prof. CNR Rao and others widespread the formulation of India's initial nanotechnology and its effective implementation by starting Nanomission programmes. India aims of manufacturing products in agriculture, textiles and energy sectors. As Prof. CNR Rao says, "India can't afford to miss the revolution in nanotechnology. We should not be at the receiving end when the world is driven by nanotechnology" [Purushotham, 2012]. The other developing nations and LDC's also possess similar interest.

South Africa has established strong collaboration with foreign partners especially Brazil and India. On the other side of the equation, India, Brazil and South Africa (IBSA) Nanotechnology Initiative aims to foster collaboration between these emerging countries. The initiative focuses on health, water treatment and agriculture; these are three areas of nanotechnology where the applications could directly bring benefits to the populations and economies of the developing world [Web: scidev. net, 2009]. An initiative of India, Brazil and South Africa promotes South-South cooperation in several arenas, including science and analysis collaboration in fields such as nanotechnology, oceanography and Antarctic investigation [Web: shirleqgy.wordpress.com, 2014]. Further, some other developing countries have joined BRICS because they have caught the vision of upcoming nanotechnology industrial revolution and have started their own nanotechnology initiatives through proper policy framework, robust budgetary plan, network linkages and human capital development for successful national development in line with the effort of Asian and Pacific Centre for Transfer of Technology - United Nations Economic and Social Commission for Asia and the Pacific (APCTT-UNESCAP) to facilitate regional collaborations in nanotechnology innovation and industrial application [Web: Nanoglobe.com, 2013]. Significantly, the Organisation for Economic Co-operation and Development (OECD) has developed questionnaire of Working Party on Nanotechnology (WPN) to collect information on overarching national or regional governmental policy and/or national or regional research programmes supporting the responsible development of nanotechnology. Delegations reported on the challenge posed by the cross-sectoral nature of nanotechnology highlighting that many government departments and agencies, with different mandates and varying levels of expertise, are affected by nanotechnology developments [Working

Party on Nanotechnology, 2013]. Figure 11.5 indicates the web of international Nanotechnology research activity [Shapira and Wang, 2010].

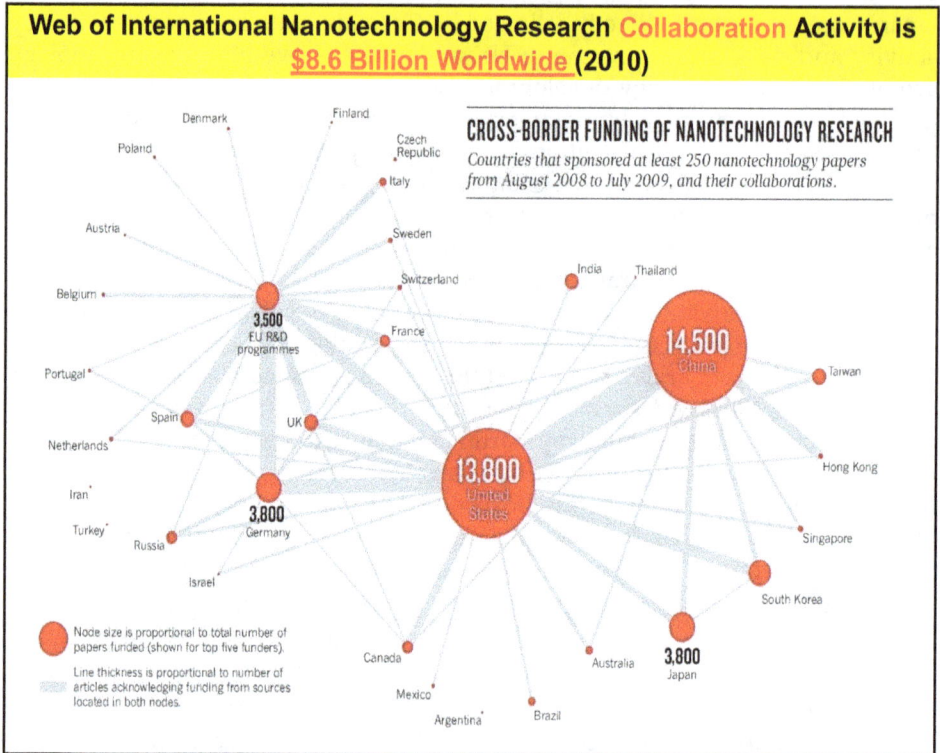

**Figure 11.5: International Nanotechnology Research Activity Web.**

## Present National Status among the Developed and Emerging Economies

Astronomer Carl Sagan has been quoted to have said that *"Somewhere, something incredible is waiting to be known"* which succinctly captures the essence of what it means to be a researcher. That wide-eyed sense that anything is possible through research and development - that we can help solve some of the most critical challenges by increasing our knowledge and understanding of the world – that is the essence of a researcher's calling [R&D Magazine/Battelle, 2013].

According to the reports of R&D levels for 2014 by www.rdmag.com, as shown in Figure 11.7 significant R&D investments by western countries in long-range technology platforms like robotics, high performance computing, social media, software, cost effective energy sources and nanobiotechnology could stimulate rapid industry-scale economic growth. Figure 11.6 displays the subject-wise ranking of global R&D leaders reflecting the significance of Composite/Nano/Advanced Materials and indicates the ranking in ascending order as the U.S. being the first

Figure 11.6: Technology Area-wise Ranking of Global R&D Leaders on Composites. Nano and Advanced Materials.

Source: Battelle, *R&D Magazine*

one among all the economically potent nations followed by Germany, Japan, China and the U.K [R&D Magazine/RDMAG, 2013].

Parametrically, various indicators are noted to study the nanotechnology ranking across the globe, *i.e.*, policy and legal framework, funding and investments, human resource development and industries scenario/economic impact. The United States has tremendous potential of nanotechnology, but nevertheless, the lead is followed by G8 countries (Japan, EU) and some developing countries (China, South Africa, Brazil, India and Thailand) that have come to this knowledge are also heavily investing. With regard to dominance, USA is a strong partner equipped with robust policies, framework and prestigious research centres and universities. Since the announcement of National Nanotechnology Initiative (NNI) by USA in 2000 almost every developed and developing economy has initiated national nanotechnology programmes. Globally, by the end of 2014 total government funding for nanotechnology worldwide will rise to $100 billion (other reports say $123 billion) and nearly a quarter of a trillion dollars will be invested into nanotechnology by 2015 [Cientifica Ltd, 2011]. In contrast, the fiscal year (FY) 2001 was commenced with a budget of $494 million but currently, FY 2014 US funding request [Roco, 2014] for nanoscale, engineering and technology R&D is $1.7 billion across 27 participating federal departments and agencies, reflecting nanotechnology potential. It is observed that the federal investment of 2014 in nanotechnology is 8.4 percent less than the actual FY 2012 budget of $1.9 billion, mainly because of the decrease in the DOD requested contribution from $426.1 million in 2012 to $216.9 million in 2014. NNI-sponsored R&D is reported in eight programme component areas (PCAs). The PCAs and proposed FY 2014 funding levels across all NNI agencies are as follows:

1. Fundamental nanoscale phenomena and processes, $445 million;
2. Nanomaterials, $368 million;
3. Nanoscale devices and systems, $400 million;
4. Instrumentation research, metrology, and standards for nanotechnology, $57 million;
5. Nanomanufacturing, $100 million;
6. Major research facilities and instrumentation acquisition, $176 million;
7. Environment, health, and safety, $121 million; and
8. Education and societal dimensions, $36 million.

While fundamental research remains the largest single NNI investment category, the research on nanodevices and systems and in nanomanufacturing would total to more than $500 million. The requested nano-EHS investment in FY 2014 is almost 10 per cent above the 2012 actual spending, without accounting for an inflation rate of four percent and the cumulative investments in nanotechnology-related environmental, health and safety research since 2005 till now is total nearly $900 million. Also for commercialisation, USA has well-established industries such as Hewlett Packard, IBM, Intel, Motorola, Texas Instrument, Dow Chemical and Raytheon which are investing heavily in nanotechnology in their collaborations with universities. The Japanese counterparts are NEC Corporation, Toshiba, Nihon

Shinku, Gijustu, NTT Docomo, Fujitsu and Sony [Oriakhi, 2004]. Venture capitalists have spent almost $900 on nanotechnology start-up firms since 1999 [Roco, 2003]. Universities have also funded nanotechnology research and development and start-up firms [Oriakhi, 2004].

EU has coordinated nanotechnology projects through Framework programmes for research and technological development incentives. The currently progressing ones are FP8 (2014-2020); Horizon 2020 (2014-2020), which is expected to focus on international corporation targets; and some other listed known working programmes, *viz.* EURAXESS and EuropeAid. Irrespective of the great economic challenges facing Europe, seven of the EU countries are actively engaged in nanotechnology activities at their national levels. Within EU8 (Germany, France, UK, Spain, Italy, Sweden, Netherlands, and Finland), in Germany for instance, nanotechnology funding stood at about 500 million euros per year. France has a budget of about 400 million euros per year with about 130 companies and over 700 nanoresearchers in nanobiotechnology. UK invests about 250 million euros per year with about 200 nanotechnology companies focusing on nanobiotechnology, nanomedicine, nanoenergy, and nanomaterials. Other countries in Europe have their investments at about 100 million euros per year and with well-tailored targets to achieve their interest and maintain global competitiveness and sustainability [Web: Observatorynano.eu]. In addition, research in the EU-27 and European Free Trade Association (EFTA) (Iceland, Liechtenstein, Norway and Switzerland) countries is closely connected, which is also mirrored in the involvement of the EFTA countries in EU research policy. While Iceland, Liechtenstein and Norway cooperate on research policy as part of the European Economic Area (EFTA, 2011), Switzerland does so via a bilateral agreement with the EU [Schiermeier, 2014]. Switzerland also contributes substantially to European research in nanotechnology and Swiss organizations are important collaborators in the European nanotechnology network. In comparison to the US and Japan, European nanotechnology activities are highly decentralized [CEC, 2007].

Considering the growing applications of nanotechnology in various fields, it is estimated that the global market for nanotechnology is expected to hit US$3.1 trillion annually from 2015, whereas, market research agency Luxresearch (www.luxresearchinc.com) [Lux Research, 2008] has estimated that the market opportunity for nanotechnology based products by 2015 would be about US$2.44 trillion. The break-up of this market opportunity is - Nano materials US$2.9 billion; Nano intermediates US$474 billion; and Nano-enabled products US$1960 billion [Lux Research, 2012]. In BRICS countries, believed to be a trillion dollar club, the market for nanotechnology is expected to be about US$1 billion [Jason, 2011]. In Russia, the focus is on using cluster manufacturing approach is to produce nanomaterials, nanomedicine, nanophotonics and nanoelectronics for ICT [Chubais, 2010]. In April 2012 they invested US $79m and Rusnano's task is to create by 2015 a nano-industry in the country that will make marketable products worth 900billion rubles ($29 billion). Alongside the BRICS nations, China is a major player in the field of nanotechnology and according to the reported 2011 estimates China has surpassed US in nanotechnology funding. In this relation, China has spent US$2.25 billion in nanotechnology research while the US invested US$2.18 billion. But China is not the first country to outspend the United States. Japan and Russia have also managed to snatch a temporary lead before falling back [Electronics.ca Research Network, 2011]. Brazil has been playing a leading role in Latin America and in 2010 the Brazilian Ministry of Science, Technology and Innovation estimated that Brazil had over 3,000 nanotechnology researchers, including professors and students; it is via the collaboration with universities that Brazilian industries are expected to reach the forefront of nanotechnology innovation. The Government of India have sanctioned Rs.1,000 crore (~US$214 million) in 2007 under Nano Mission [GOI, 2007], an umbrella programme to promote R&D in nanotechnology and efforts are put forward to forge international collaborations. The Government has now approved US$109.42 million for Nano mission in its second phase in the 12th Plan Period (2012-17) [Mohan, 2014].

Thus a clear trend is emerging: while nanotechnology research spending in Europe and North America is still rising, the fast growth rates are seen in Asia. Asian (comprising 49 countries) investment in nanotechnologies was poised to be the largest in the world until RusNano was formed with its huge budget. For example, Israel invested $101 million in last 5 years on nanotechnology [NanoIsrael, 2012]. Thailand's programme receives approximately US$2 million per year in which the success recorded so far is by collaborative networks in research and funding by various government agencies through their various COEs in nanotechnology [Maclurcan, 2005]. Similarly, Singapore has an elaborate nanotechnology capabilities utilising nanomaterials, nanodevices in microelectronics/MEMS fabrications, clean energy and medical technology, among others, in so many well-established nano-SMEs involving technology/manufacture and sales/marketing under government funding and collaborative arrangements [Lerwen, 2010]. However, Bangladesh and Nepal [Nanoglobe, 2009] [APCTT-UNESCAP, 2010] have not launched nanotechnology initiatives due to their limited infrastructure for R&D, lack of trained human resources and limited international collaboration. Further, some key nanotechnology areas that are thought to be of most potential benefit for the Asia-Pacific Region includes information and communication technology (miniaturization and efficient material development); healthcare (diagnostic, cancer treatment, and biosensors); environmental protection (reduce carbon dioxide emissions); reduction of energy consumption; purification, protection, and production of drinking water (arsenic mitigation and nanofiltration); renewable energies; and agriculture and food security/safety.

In the Middle East the regional leader in nanotechnology is Iran [INIC – News, 2010] [Bernama, 2009] [Web: Wayback.archive.org.]. In order to be influential among Islamic developing countries, Iran founded the Organisation of Islamic Cooperation's Standing Committee on Scientific and Technological Cooperation (OIC-COMSTECH) with the help of the World Academy of Science (TWAS) and in the Joint Research Grants Programme, Iran pushes forward the dialogue on nanotechnology. The country also plays an important role in the Nano Technology Network of the Organization of the Islamic Conference (NTNOIC), a global discussion area for nanotechnologies [TWAS-COMSTECH, 2010]. Moreover, Iran had its National Nanotechnology Initiative launched in 2005 for a 10-year period up to 2015 with broad mark achievements and half of its nanotechnology budget is funded by the private sector with her scientists and industries actively engaging in international cooperation activities.

In Africa, R&D has been led by an emerging South Africa, which after 2 years of NNI launched its support for nanotechnologies with its South African Nanotechnology Initiative. The South African strategy focuses on health, water and energy as a first pillar followed by the chemical and mining industry, thus acting as the driving force of nanotechnologies in Africa. Further new nanotechnology initiatives are also launched by Nigeria, Zimbabwe and Kenya. At the regional level Africa currently hosts a couple of initiatives, enumerating, the Nanosciences African Network (NANOAFNET) [Web: nanoafnet.tlabs.ac.za]; African Organisation Focus Nanotechnology Africa Inc. (FONAI) [Web: fonai.org]; African Network for Solar Energy (ANSOLE) in which Austrian authorities are also involved, etc [Web: ansole. org/]. It is observed that technological proximity has effect on policy measures, in particular the financial incentives, are strong enough management that will take this diverse approach on collaboration on board but unfortunately, most African nations and some other least developed countries (LDC) have only demonstrated interest to start without any practical approach to its implementation. Therefore a strong collaboration link between African nations and nations like South Africa, India, and European Union with strong nanotechnology capabilities should be established. At the same time it can be noted that US has maintained a competitive position in the worldwide nanotechnology marketplace while realising nanotechnology's full potential and creating a strong response from the rest of the world, placing nanotechnology as a priority area in their S&T policy.

For an overall comparison, Table 11.2 illustrates a more comprehensive data for Russia, US and China on Nanotechnology market [World Bank, Rosstat, Forfas (2010), Roco (2010), Shapira and Wang (2010), Kachak *et al.* (2010)].

**Table 11.2: Overall Nanotechnology Status in Russia, USA and China**

| | Russia | USA | China |
|---|---|---|---|
| GDP (billions, constant 2000 US$) | 397,95 | 11250,7 | 2937,55 |
| GDP per capita (constant 2000 US$) | 2805 | 37016 | 2206 |
| Gross R&D expenditure (as % of GDP) | 1,24 | 2,67** | 1,49** |
| Gross nanotechnology-related R&D expenditure (millions US$) | 504 | 3700* | |
| High-technology exports (billions, current US$) | 5,11* | 231,13* | 381,35* |
| Patent applications (residents) | 27712* | 231599* | 194579* |
| Nanotechnology patents issued | 338 | 6729* | |
| Researchers per million population | 2602 | 4663*** | 1071** |
| R&D personnel in the sector of nanoscience and nanotechnology | 14500 | ~150000* | |
| Number of nanotechnology publications (08.2008-07.2009) | ~2700 | ~21000 | ~20100 |
| Domestic market for nanotechnology products (billions US$) | 2,7 | 80* | |

* Data available for 2008. ** Data available for 2007. *** Data available for 2006.

## Knowledge and Technology Transfer: Trends in Co-publication and Co-patenting

The OECD (2003) has defined technology transfer (TT) as IP management, focussing on patenting and licensing – services frequently provided by technology transfer offices (TTOs). Knowledge transfer is a broader concept that also encompasses TT, acknowledging the many forms, activities, processes, and actors involved in making knowledge from the research sector available for creating benefits throughout society. There is clearly agreement now that knowledge transfer is broader than IP management. There is also a wide array of approaches to conceptualising, classifying, and measuring it [Link, 2003].

In connection, strong technological capabilities, commercialization of research, and ability to acquire and adopt technologies from internal and external sources are of key importance for the successful economic growth of a country. Therefore, Technology transfer is an important area that needs S&T diplomacy.

The growing competitive performances in nanotechnology R&D within the nations have revealed evolutionary trends and characteristics of intensification of international collaboration. It is indicated by co-publications and co-patenting at the global scale to benefit and profit companies, both in basic and applied research, thus building and strengthening their internal technological capabilities. The multiple interaction channels enable companies to cut across established disciplinary and sectoral boundaries. It is evident that the international collaboration publication quality solely focuses on metric values measured by different patterns, namely, combined bibliometrics and science mapping, co-authorship analysis and citation networks of papers. Owing to the rapid development and ardent enthusiasm for nanotechnology globally, the scientific publications produced by Israel, the Netherlands, the US, Germany, Austria and Switzerland have the highest scientific impact, while nanotechnology development in this sector is more intensive outside Europe, Japan, The US and the South Korea [Explanatory Memorandum on European Union Document, 2013]. Apparently, it is observed that in international collaborations physical sciences has the highest publication impact while in regional collaboration life sciences have higher citation impact [Web: ip-science.thomsonreuters.com]. As a whole, ERA countries, EU member states have significant scientific impact in ICT. European science has, however, higher quality in publications related to industrial sectors, such as aeronautics, new production technologies, other transport technologies and materials [Marcus, 2013]. However, developing countries have a very low number of scientific publications published and they could act as regional drivers, thus delivering appropriate assistance to LDC's to gain access to nanotechnologies. As a matter of fact, China leads in production of nanoscience followed by USA but the case is vice-versa otherwise considering the international collaboration rate.

While studying co-patenting [Zheng, 2013], it is found that international collaboration nanotechnology patents across the world has likewise showed continued growth until 2006 but the results in number have declined; the time lag between patent application year and patent approval is one such factor. Patents provide a reliable quantisation basis for technology collaboration studies, with an aim to contribute to policy makers and relevant managers when making decisions for university, firm locality and choices on collaborators. It is worth mentioning that the gap between international collaboration nanotechnology patents and total nanotechnology patents in terms of citation impacts had become narrower; the gap from 1991 to 1997 was 0.77, 1998 to 2003 was 0.15, and 2004 to 2010 was 0.06. The proportion of international collaboration nanotechnology patents in the total nanotechnology patents has increased from that of less than 5 per cent in 1991 to more than 9 per cent in 2010. Graphically, USA took the leading position accounting for over one-third of the international collaboration nanotechnology patents in the world. The percentage of the international collaboration patents over the total number of patents remains relatively low, *i.e.* 10.4 per cent due to large number of total nanotechnology patents; and Japan, South Korea and Taiwan with share lower than 15 per cent due to maintaining high numbers of international collaboration nanotechnology patents. The proportion of international nanotechnology co-patenting is 40 per cent for Switzerland, the UK, the Netherlands, Belgium, Sweden, Russia, India and Singapore. The number of these countries remains low when considering the domestic nanotechnology patents followed by the rankings of Germany, France, Canada, China, Israel, Australia and Italy. Switzerland has the highest rate of all countries with 63.5 per cent of international nanotechnology co-patents. Asian countries/regions have shown an obvious increase in the number of nanotechnology collaboration patents. Studies also revealed that the same country may adopt different collaboration activity pattern in publication or patent area. For example, Japan cooperated with a wide range of countries in patenting activities, but only selectively cooperated with some leading countries in publications.

The total number of nanotechnology patents over the past 20 years is 17,899, which has exhibited continuous growth from 1990 to 2006, as illustrated in Figure 11.7. Since 2006, the number of patents has gradually declined. The time lag between the application year and patent approval year may have been one factor that has resulted in the apparent decline in the number of patents since 2006, as data were collected and analysed based on patent application year. International collaboration nanotechnology patents across the world has likewise showed continued growth until 2006, with the number increasing from 5 in 1991 to 163 in 2006. The time lag between patent application year and patent approval may also be the factor that has resulted in the decline of international collaboration patents after 2006 on the database [Zheng, 2013].

## Challenges of Nanotechnology Development

There are various upcoming challenges of nanotechnology in developing countries and LDC because of fund scarcity and political issues that blind them against realities of life. In medium and low income countries, economies are mostly without a strong science base that lacks government R&D funding; proper legislation/regulatory framework; infrastructure; human resource and policy capacity; ability to translate R&D investments into economic outcomes and inadequate foreign linkages. For instance, Nigeria and most other African countries lack the basic materials characterisation equipment, while Bangladesh

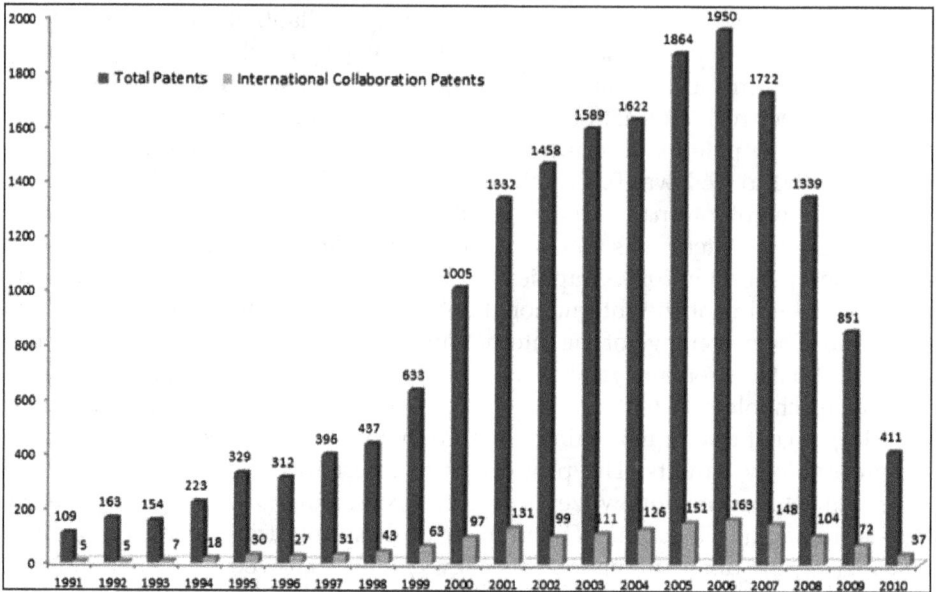

**Figure 11.7: The Number of Total Patents and International Collaboration Patents in Nanotechnology by Year.**

and Nepal have not launched nanotechnology initiative so far. Moreover, lack of R&D partnerships among the countries is also a major issue in South Asia.

Overall, the challenges are listed as under:

☆ Lower government spending on R&D

☆ Lack of infrastructure and human capacity

☆ Lack of proper education relating to curriculum development matters

☆ Lack of proper legislation/regulatory framework and the relevant political drive

☆ Lack of private enterprise participation in R&D

☆ Lack of proper collaboration and network programmes among agencies

☆ Research institutes and industries to translate basic research into applied research and end products

☆ Poor industrialisation status of the third world countries

☆ Inadequate foreign linkages, particularly with donor agencies in nanotechnology

☆ Fear of health, environmental, and safety risks associated with nanotechnology

Globally, on the surface of the modern business model, the protection of intellectual property is a critical factor shaping international competition, real economic growth and wealth creation. Moreover, the major challenge of nanotechnology development is converting R&D to commercial products and its applications.

Herein is the paradox. It is important to identify the key issues of nanotechnology national and global collaborative networks that are driving their trajectory globally. Many questions are lying beneath the collaborative projects, such as, whether greater collaboration will emerge, or at the other extreme, we will see the ultimate breakdown; which countries will manufacture and which will become nanotechnology importers; does this new subject have relevance to developing countries agenda. In this context, developing countries and several commentators (mainly based in western countries) are progressing at a steady pace raising a major issue to developing countries that nanotechnology can aggravate global inequalities. This new technological wave can gradually strike deepening, rather than helping to bridge the global divide, indeed, 'nano divide' [Brahic, 2004] [Maclurcan, 2009]. Monopolies, concentrated control and ownership penetrate political systems by creating management uncertainties. Negatively phrased, among the BRICS countries advancements in nanotechnology made in China and Russia are enormous that they are no longer in the same categories with other member nations hence their inclusion is regarded as national activity nations [Ezema, 2014]. Some criticism centres on the assertion that the G8 members do not do enough to help global problems such as developing countries debt, global warming and the AIDS epidemic due to strict medicine patent policy and other issues related to globalisation [Alexander, 1996]. At the EU level, member states are at different stages of nanotechnology development [Web: en.wikipedia.org/wiki/G8#cite_note-95]. This is further linked to fragmentation of activities in terms of funding opportunities and safety discussions being a significant barrier for a common European approach.

Concerns are also expressed that the nature of global risks is constantly changing. Thirty years ago, Chlorofluorocarbons (CFCs) were seen as a planetary risk. Then the proliferation of nuclear weapons occupied the minds of scientists and politicians. Recently it was asbestos and today the issues are centred on Carbon Nanotubes (CNTs).

Measurement standards by various technical committees (TC) are established globally to deal with standardising issues. Besides Nanotechnology TC's, the two major globally-relevant standards development organisations ISO and IEC have been developing standards to facilitate the safe and responsible use of nanotechnologies since 2005. The specific Technical Committees - ISO/TC229 and IEC/TC113 - have over 30 participating member countries. ISO/TC229, for example, has the following objectives through international standards development [Nanotechnologies Standards Development List, 2012]:

- ☆ To support the sustainable and responsible development of nanotechnologies
- ☆ To facilitate global trade in nanotechnology-enabled products and systems
- ☆ To improve quality, safety, security, consumer and environmental protection, together with the rational use of natural resources
- ☆ To promote good practice in the production, use and disposal of

nanomaterials and nano-enabled products.

At the Technical Committee (TC) level, ISO/TC229 and IEC/TC113 are respectively chaired by the UK and the USA. For operating efficiency, the work of the TCs is subdivided into working groups. Each working group has a designated country that has volunteered to convene its activities, and within each of these groups, work items are initiated following a specific standards development process with an end deliverable as a technical report, technical specification, or international standard in accordance with a TC-directed business plan.

## Planning Strategies of Developing Nations

Nanotechnology has been stimulated by three factors. These are science, technology as well as policies associated with innovation. Prominently, *it is recently estimated that governments around the world invested $67 billion* over the last **11 years** [Web: cientifica.com, 2011]. In a majority of discussions, strategies for the responsible development of nanotechnology are based on cross-ministry/cross-departmental involvement.

It is identified that India and China commonly use three unique strategies [Ruan, ETM/IEL Working Paper], namely;

☆ Frugal Engineering: Innovators adapt more expensive technology to lower cost solutions that can be marketed to large numbers of people with lower incomes.

☆ Modularisation: This refers to building a product on a modular architecture and working with the suppliers towards standardised modules so as to achieve the economy of scale based on specialisation. Generally, for example, Tata Nano Car is designed on an open structure involving part suppliers from all over the world. By setting a price limit at the upfront, letting suppliers bid for the projects and manufacture modularised parts, and then having the local manufactures assemble the cars, Tata significantly reduced the total cost. Another example is the China's E-bike industry. The companies were able to purchase modularised parts in the market and put together cheap but good enough E-bikes which slowly became substitutes for bikes and motorcycles for the Chinese customers.

☆ Drastic Reduction in Manufacturing Cost: This refers to a collection of re-engineering processes which replaces sophisticated expensive machinery with self-invented procedures, expensive raw material/parts with slightly lower performance but much lower cost, and fast expansion of business to push the cost further down.

Other nations also focus on augmentation strategy where firms can augment an existing product with a disruptive feature.

China has also fixed a macroeconomic goal of spending 2.2 per cent of GDP on research by 2015, toward becoming an innovation-based economy by 2020. Such a command approach can sometimes accelerate the translation from research to development. This is illustrated in Figure 11.8 [R&D Magazine/Battelle, 2013], showing large proportion of development investment in China versus funding for basic and applied research. It is manifested (for example) in the large-scale deployment of clean energy and advanced grid technologies in China.

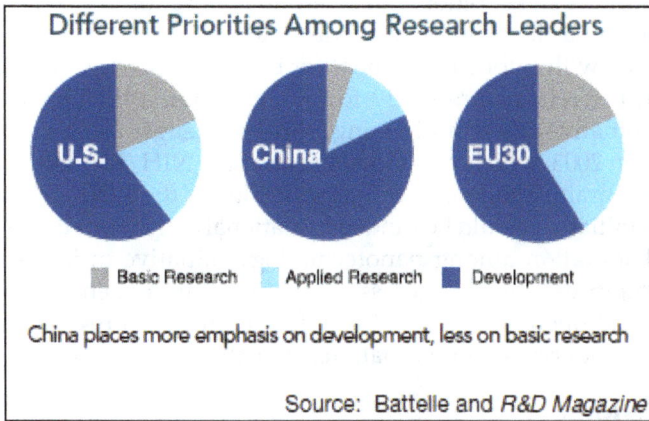

**Figure 11.8: Investment Strategy in EU, USA and China.**

For the development of a goal normative, the active participation in nanotechnology collaboration of global integration is to be real. The point is to infuse new technology in the market and facilitate linkages with robust regulations, agreements and channelised funding into concerned sector. Especially in the fields of water, energy, and health, it has been pointed out that nanotechnology can contribute to the creation of cheaper and more efficient technologies that can help the poor, such as improved water filters, energy storage systems, solar powered electricity and portable diagnostic. Collaborations can fill up the nano divide gap, which should take place, for example, in Nanomission Programme of India, NNI of US, FP8 of EU and Rusnano of Russia. Short and long term plans on nanotechnology should be set. In developing nations, the 'brain-drain' can be reversed and the skills obtained by the researchers who have worked abroad, must be seen as asset to spur on the development of nanotechnologies in these countries. LDC and African nations should review there tertiary education programmes, not by promoting talks about 'nano divide', from which they will suffer more as consuming nations. Furthermore, though it is useful to keep in mind some notes on the US system of innovation, the follower countries should restrain themselves from fully imitating the US model and strategies. It is rather imperative for them to focus on the concentrated national technological needs.

The planning strategies should also involve some specific points mentioned below:

☆ Setting up of new research units in established centres of excellence

☆ Financing of specific research projects

☆ Building human resource capabilities

☆ Initiating public-private partnerships

☆ Gearing up the regulatory framework: Institutional mandates and issues of synergy

☆ Facilitating technology transfer

In addition, the stakeholders must zero in on a set of 'concrete targets' to be accomplished similar to in 'mission mode', like for space technology. By playing a mixed strategy with public investment for both economic growth and inclusive development, the returns are likely to be higher. On a broader perspective, a Global Fund initiative called 'addressing global challenges using nanotechnology' was proposed in 2003, which was modelled after the NIH/Bill and Melinda Gates Foundation for grand challenges in global health in the LDC's. [Varmus, 2003]. Moreover, the initiative would be funded by national and international foundations and from collaboration among nanotechnology initiative in industrialised and developing countries. This will access innovations through encouraging national governments and open access to publicly funded research results and materials, thus benefiting both rich and poor nations at relatively low cost.

## 2. Conclusions

The knowledge-based economy paradigm and the orientation of S&T toward increasing international competitiveness were the theoretical frameworks that justified the designation of nanotechnology as priority area, even in the emerging countries. At this moment it is more exploratory, but nano-, bio- and information technology areas are expected to grow in strong synergism with cognitive sciences. There is a convergence of sciences in the 'nanoworld'. Nanoscale science and engineering R&D is mostly in a growing and pre-competitive phase. International collaboration in fundamental research, long-term technical challenges, metrology, education and studies on societal implications will play an important role in the affirmation and growth of the field. The vision setting and collaborative model of National Nanotechnology Initiative has received international acceptance. Most industrialised nations are establishing or are planning to establish their national programmes. Enhancing communication, networking for exchanges of peoples and ideas and development of R&D partnerships are sought for added value in research and leveraging.

Priority science and technology goals may be envisioned for international collaboration in nanoscale research and education through better comprehension of nature, increasing productivity, sustainable development and addressing of humanity and civilisation issues. Moreover, the planning strategies to foster research in nanotechnology vary from nation to nation. It is imperative to note that the current economic input via public investments and by other major funding agencies across the world is certainly not equal to the output in the form of application sector of nanotechnology. However, the focus is on the input driven science to output driven science and hence there is paradigm shift towards the change in drift of science policy. Therefore, the world is on the outlook for innovations that are investable and aims for creating such a platform that leverages technology advantage which can be permeated to a large market. To expertise developments, federal, state and local government level should be mobilised to enter into collaborations, in particular, through North-South and South-South cooperation. In conclusion, Nanotechnology is still in its infancy and will take time to deliver on its promises. The developing world will also need time to appropriate the technology so as to make the most out of it and to boost its economies.

## 3. Acknowledgement

My special gratitude to Prof. Dr. Arun P. Kulshreshtha, Director General and Mr. M. Bandyopadhyay, Senior Expert and Administrative Officer of NAM S&T Centre for giving me the opportunity for my participation in this workshop. I also wish to thank the staff of the NAM S&T Centre for their overall support.

## REFERENCES

1.  <http://en.wikipedia.org/wiki/G8#cite_note-95>

2.  <http://ip-science.thomsonreuters.com>

3.  Alexander T. (1996). Unravelling Global Apartheid: an overview of world politics. Polity Press. pp. 212–213.

4.  Ansole. <http://www.ansole.org/>

5.  APCTT-UNESCAP. (2010). Proceedings and Papers Presented at the Consultative Workshop on Promoting Innovation in Nanotechnology and Fostering Industrial Application: an Asia-Pacific Perspective. Innovation in nanotechnology: an Asia-Pacific perspective. <http://www.nis.apctt.org/PDF/Nanotech_Report_Final.pdf>

6.  Bernama. (2009). Iran Ranks 15th in Nanotech Articles.

7.  Brahic C. (2004). Science of the small could create nano-divide. SciDev. net. <http://www.scidev.net/en/news/science-of-the-small-could-create-nanodivide.html>

8.  CEC (Commission of the European Communities). (2007). Nanosciences and Nanotechnologies: An Action plan for Europe 2005-2009. First Implementation Report, 2007-2009, Communication from the Commission to the Council, the European Parliament and the European Economic and Social Committee, Brussels, 29.10.2009, COM(2007) 505, mimeo, <ftp://ftp.cordis.europa.eu/pub/nanotechnology/docs/com_2007_0505_f_en.pdf>

9.  Chubais A. (2010). USRBC 18th Annual Meeting "From Silicon Valley to Skolkovo: Forging Innovation Partnerships". RUSNANO: Fostering Innovations in Russia through Nanotechnology. <http://www.usrbc.org/pics/file/AM/2010/./chubais_GB_830.ppt.pptx>

10. Cientifica Ltd. (2011). The economic impact of nanotechnologies.

11. Cientifica Ltd. (2011). Global Funding Of Nanotechnologies and Its Impact.

12. Cientifica.com (2011). Global Funding of Nanotechnologies. <http://www.cientifica.com/research/market-reports/nanotech-funding-2011/>

13. Clinton W. (2000). Presendent Clinton's Address to Caltech on Science and Technology. <http://pr.caltech.edu/events/presidential_speech/pspeechtxt.html>

14. Cunningham S.W. and Werker C. (2012). Proximity and collaboration in European nanotechnology. Papers in Regional Science, Wiley Blackwell, vol. 91(4), pages 723-742, November.

15. Drexler K.E. (1986). Engines of Creation: The Coming Era of Nanotechnology. Doubleday. ISBN 0-385-19973-2.

16. Electronics.ca Research Network. (2011). Annual Global Nanotechnology Research Funding Running at $10 Billion Per Year. Semiconductor Research News. <http://www.electronics.ca/presscenter>

17. Explanatory Memorandum on European Union Document. (2013). Commission Staff Working Document 'Innovation Union Competitiveness report'. Submitted by Department for Business Innovation and Skills on 20 February 2014.

18. Ezema I.C., Ogbobe P.O. and Omah A.D. Initiatives and strategies for development of nanotechnology in nations: a lesson for Africa and other least developed countries. Nanoscale Research Letters, 9:133, doi:10.1186/1556-276X-9-133.

19. Fonai. <http://fonai.org/Home_Page.php>

20. Forfas. (2010). Ireland's Nanotechnology Commercialisation Framework 2010–2014. <http://www.forfas.ie/media/forfas310810-nanotech_commercialisation_framework_ 2010-2014.pdf>

21. Global List of Organizations and Efforts Related to Nanotechnology, Nanoscience, Nanomaterials, and Food and Agriculture Products. Food and Chemical Safety Committee (2009).

22. http://www.ilsi.org/NorthAmerica/Documents/FOOD per cent 20CHEMICAL per cent 20SAFETY/Global per cent 20List per cent 20of per cent 20Organizations per cent 20and per cent 20Efforts per cent 20Related per cent 20to per cent 20Nanotechnology.pdf>

23. Global Nanotechnology Funding and Impacts. www.luxresearchinc.com. (2012)

24. GOI. (2007). 11th Five-Year Plan 2007-2012. Planning Commission, Government of India. New Delhi: SAGE Publications India Pvt Ltd.

25. INIC – News. (2010). 73 per cent of Tehran's Students Acquainted with Nanotechnology. <En.nano.ir.>

26. International Engagement. United States National Nanotechnology Initiative. <http://www.nano.gov/initiatives/international>

27. Jason. (2011). Global Science Research and Collaboration. <http://www.globalsherpa.org/research-science-technology-international>

28. Kachak V.V. (2010). About realization in 2009 of the Program of Nanoindustry Development in the Russian Federation until 2015 (in Russian). Russian Nanotechnology 5(9-10), 11-13.

29. Lerwen L.I.U. (2010). Singapore nanotechnology capabilities report. NanoGlobe Pte Ltd. <https://www.engineersaustralia.org.au/./Nanotechnology per cent 20capabilities>

30. Link, A. N. and K. R. Link. (2003). On the growth of U.S. science parks. Journal of Technology Transfer 28:81–85.

31. Lux Research. (2008). Overhyped technology starts to reach potential: nanotech to impact $3.1 trillion in manufactured goods in 2015. <http://www.luxresearchinc.com/press/RELEASE_Nano-SMR_7_22_08.pdf>

32. Maclurcan D.C. (2005). Nanotechnology and developing countries – part 2: what realities. <http://www.azonano.com/article.aspx?ArticleID=1429>

33. Maclurcan D.C. (2009). Southern roles in global nanotechnology innovation: Perspectives from Thailand and Australia. Nanoethics, 3:137-56.

34. Marcus C., Stierna J. and Moise C. (2013). Co-development of Science and Technology at a National Level and the Use of European Funding Instruments; Innovation Union Competitiveness papers. DG Research and Innovation, Economic Analysis Unit.

35. Mohan V. (2014). Govt approves Rs 650 crore for Nano mission. <http://timesofindia.indiatimes.com/>

36. Nanoglobe (2009). Nanotechnology initiatives/programs in Iran, Pakistan, Philippines, Sri Lanka and other developing countries in the Asia Pacific Region. Highlights of the United Nation APCTTESCAP Consultative Workshop on Promoting Innovation in Nanotechnology and Fostering its Industrial Application: an Asia–Pacific Perspective. <http://www.nano-globe.biz/News/UNNanoColomboDec09.pdf>

37. NanoIsrael (2012). 5 Years of Nano-Technology as an Israeli National Project. <http://www.businesswire.com/news>

38. Nanoscience African Network Mission. <http://www.nanoafnet.tlabs.ac.za/>

39. Nanotechnologies Standards Development List. (2012). CSA Group. <http://nanoontario.ca/wp/wp-content/uploads/2012/07/Nano-Standards-Development-List.pdf>

40. Nanotechnology Center for Learning and Teaching (NCLT). (2009) <http://community.nsee.us/index.php?option=com_content and view=category and id=89:international-nanotechnology-initiatives and layout=blog and Itemid=89>

41. Observatory NANO. Public Funding of Nanotechnology, Seventh Framework Programme. <http://www.observatorynano.eu/publicfunding ofnanotechnologies>

42. Ogawa H., Nishikawa M. and Abe M. (1982). Hall measurement studies and an electrical conduction model of tin oxide ultrafine particle films. Journal of Applied Physics 53, 4448.

43. Oriakhi C. (2004). Commercialization of nanotechnologies. M.S. Thesis, Massachusetts Institute of Technology, Sloan School of Management, Management of Technology Program.

44. Purushotham H. (2012).Transfer of nanotechnologies from R&D institutions to SMEs in India: Opportunities and Challenges. Tech Monitor. Oct-Dec 2012.

45. R&D Magazine/Battelle. (2013). 2014 R&D Magazine Global Funding Forecast Executive Summary.

46. R&D Magazine. www.rdmag.com. (2013). 2014 Global R&D Funding Forecast.

47. Rensselaer. (2004). Efficient filters produced from carbon nano tubes through Rensselaer Polytechnic Institute-Banaras Hindu University collaborative research.

48. Roco and Bainbridge. (2001). Societal Implications of Nanoscience and Nanotechnology, Springer, new updated 2 vols., 2007.

49. Roco M.C. (2003). Broader societal issues of nanotechnology. J Nanoparticle Res 5: 181-9.

50. Roco M.C. (2004). Nanoscale science and engineering: unifying and transforming tools. AIChE J 50(5):890–897.

51. Roco M.C. (2011). The long view of nanotechnology development: the National Nanotechnology Initiative at 10 years. J Nanopart Res. 13:427–445 DOI 10.1007/s11051-010-0192-z.

52. Roco M.C. (2014). National Nanotechnology Investment in the FY 2014 Budget. <http://www.aaas.org/sites/default/files/migrate/uploads/14pch23.pdf>

53. Roco, M.C. (2010). The long view of nanotechnology development: The National Nanotechnology Initiative at ten years. In: World Technology Evaluation Center. Nanotechnology long-term impacts and research directions: 2000-2020. Springer, 31-53.

54. Rosstat (Federal State Statistics Service of the Russian Federation). (2010). Russia's Statistical Yearbook 2010 (in Russian). Moscow: Rosstat.

55. RS Policy document (2010). New Frontiers in Science Diplomacy. ISBN: 978-0-85403-811-4

56. Ruan Y., Hang C.C. and Subramanian A.M. ETM/IEL Working Paper. Disruptive Innovation in Emerging Markets: Strategies Used in India and China. <http://www.eng.nus.edu.sg/etm/research/publications/iel1201.pdf>

57. RUSNANO (Russian Corporation of Nanotechnologies). Business Strategy until the Year 2020. <http://www.rusnano.com/Admin/Files/FileDownload.aspx?id=1772S>

58. Schiermeier Q. (2014). EU–Swiss research on shaky ground. Nature 506, 277. doi:10.1038/506277a

59. Scidev.net. (2009) <http://www.scidev.net/global/link/india-brazil-south-africa-nanotechnology-initiativ.html>

60. Shapira P. and Wang J. (2010). Follow the money. Nature 7324(468), 627-628. doi:10.1038/468627a

61. Shirleqgy. (2014) <https://shirleqgy.wordpress.com/2014/06/>

62. The Royal Society and the Royal Academy of Engineering: Nanoscience and nanotechnologies: opportunities and uncertainties. (2004). Full report.

63. The World Academy of Sciences. (2010). TWAS-COMSTECH Joint Research Grants. <http://twas.ictp.it/prog/grants/twas-comstech-joint-research-grants>

64. UN-REDD. (2011). The United Nations Collaborative Program on Reducing Emissions from Deforestation and Forest Degradation in Developing Countries. <http://www.un-redd.org/>

65. Varmus H., Klausner R., Zerhouni E., Acharya T. and Daar A.S. (2003). Grand challenges in global health. Science. 2003;302:398–399]

66. Wayback.archive.org. Iran daily: Iranian Technology from Foreign Perspective.

67. Working Party on Nanotechnology (2013). Responsible Development of Nanotechnology, Summary Results from a Survey Activity. DSTI/STP/NANO(2013)9/FINAL.

68. Zheng J., Zhao Z-y, Zhang X, Chen D-z and Huang M-h. (2013). International collaboration development in nanotechnology: a perspective of patent network analysis. Scientometrics DOI 10.1007/s11192-013-1081-x

## Chapter 12

# Specialized Centre for Scientific Research and Treatment with Laser

*Ihsan, Fathallah Rostum*

**Vice President**
**Iraqi Laser Society**
*e-mail: nouralihsan@yahoo.com*

## What is the Laser?

The term *laser* is an acronym composed of the first letters of the words *light amplification by stimulated emission of radiation*. Of these, the most important is *radiation;* the other words describe the means by which lasers generate radiation. *Radiation* may be defined as the transmission of energy from one point in space to another, with or without an intervening material medium. *Electromagnetic radiation* requires no medium for its transmission; it can travel through free space devoid of any matter. It can also be propagated through space containing matter in the form of gases, liquids, or solids. Upon entering such media from free space, electromagnetic radiation will, in general, be changed in direction and speed of propagation [1].

Electromagnetic spectrum also includes radio waves, microwaves, infrared radiation, visible light, UV - rays, x - rays, and gamma rays, all these forms of electromagnetic radiation are fundamentally similar, in that they travel at the speed of light (186,000 miles/s). The difference between them is their wavelength and energy; the shorter the wavelength, the higher the energy. For example, Radio waves have the longest wavelength (tens to hundreds of meters) and the lowest energy, Figure 12.1 [2].

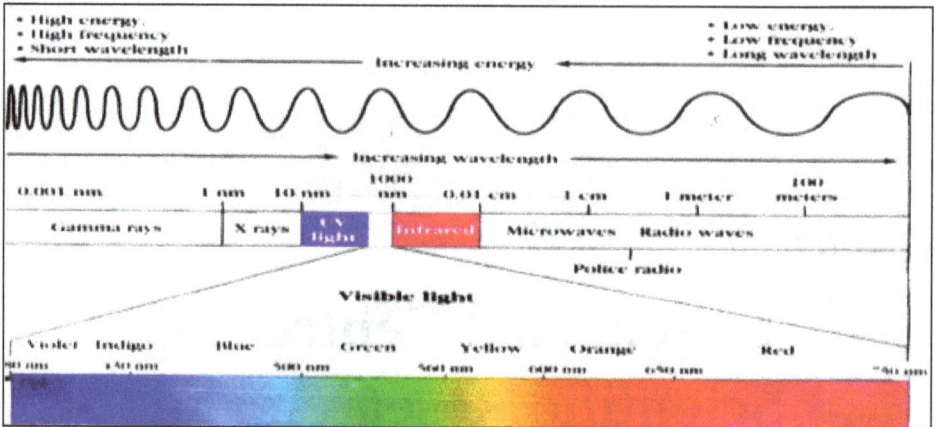

**Figure 12.1: Different Electromagnetic Wavelengths. cited by [3].**
**Less than 400 nm = ultraviolet spectrum**
**between 400 - 700 nm = Visible light**
**700 - 100 000 nm = infrared spectrum.**

## History of Laser

*Laser* can be easily explained by understanding that light is a form of electromagnetic radiation. Max Planck, Figure 12.2, received the Nobel Prize in physics in 1918 for his discovery of elementary energy quanta. Planck explained the relationship between energy and the frequency of radiation, essentially saying that energy could be emitted or absorbed only in discrete chunks – which he called quanta

In 1905, Einstein, Figure 12.3, proposed that light delivers its energy in chunks, in this case discrete quantum particles now called photons.

**Figure 12.2: Max Planck.**

In 1917, Einstein proposed the process which he called stimulated emission. He theorized that, besides absorbing and emitting light spontaneously, electrons could be stimulated to emit light of a particular wavelength.

In 1951: Charles Hard Townes of Columbia University in New York, Figure 12.4, conceives his maser (microwave amplification by stimulated emission of radiation) idea while sitting on a park bench in Washington, after that and on 1954, he worked with

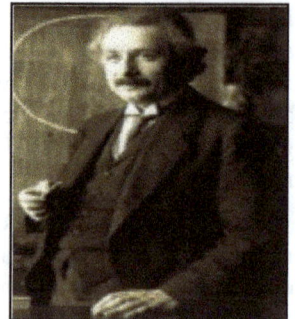

**Figure 12.3: Einstein.**

Herbert J. Zeiger and graduate student James P. Gordon. Townes demonstrates the first maser at Columbia University, The Ammonia Maser.

In 1955 and at Lebedev Physical Institute in Moscow, Nikolai G. Basov and Alexander M. Prokhorov attempted to design and build oscillators. They propose a method for the production of a negative absorption that was called the pumping method.

In 1958: Townes, a consultant for Bell Labs, and his brother Arthur L. Schawlow showed that masers could be made to operate in the optical and infrared regions and propose how it could be accomplished. At Lebedev Institute, Basov and Prokhorov also are exploring the possibilities of applying maser principles in the optical region in 1955.

**Figure 12.4: C. Townes.**

In March 22, 1960: Townes and Schawlow are granted US patent number 2,929,922 for the optical maser, now called a laser.

In May 16, 1960: Theodore H. Maiman, a physicist at Hughes Research Laboratories in Malibu, Calif., constructs the first laser using a cylinder of synthetic ruby measuring 1 cm in diameter and 2 cm long, with the ends silver-coated to make them reflective and able to serve as a Fabry-Perot resonator. Maiman uses photographic flash-lamps as the laser's pump source.

In December 1961: The first medical treatment using a laser on a human patient is performed by Dr. Charles J. Campbell of the Institute of Ophthalmology at Columbia-Presbyterian Medical Center and Charles J. Koester of the American Optical Co. at Columbia-Presbyterian Hospital in Manhattan. An American Optical ruby laser is used to destroy a retinal tumor. In 1962: With Fred J. McClung, Hellwarth proved his laser theory, generating peak powers 100 times that of ordinary ruby lasers by using electrically switched Kerr cell shutters. The giant pulse formation technique is dubbed Q-switching. Important first applications include the welding of springs for watches.

In 1962: Groups at Lincoln Laboratory simultaneously develop a gallium-arsenide laser, a semiconductor device that converts electrical energy directly into infrared light but which must be cryogenically cooled, even for pulsed operation [4].

## What's make Laser so Unique Tool?

Laser radiation is characterized by an extremely high degree of Monochromaticity, Coherence, Directionality and High Intensity.

### 1. Monochromaticity

The light from a laser is monochromatic, which means that it is of a particular wavelength, or of a single color. Light from sodium lamp is monochromatic, *i.e.* of single color or of single wavelength of about 58930A.

On the other hand, the band width of an ordinary laser lies at the order of $10^{\circ}A$ while a high quality one lies in $10^{-8}$ at $6000^{\circ}A$. This narrow band width of a laser light is called *'high monochromacity'*. Because of this monochromaticity, large energy can be concentrated into an extremely small band width.

## 2. Coherence

The light from a laser is coherent, which means that each photon is in synchrony with the other photons, or the patterns of their waves are aligned with each other, thus increasing the intensity of the light emitted.

Visible light energy is emitted when the excited electrons in atoms undergo transitions to the ground state. In ordinary light sources, these transitions take place at random in time, thus the light waves received at a point on a screen bear no definite phase relation among them. But, in a laser source, electronic transitions take place in an 'orderly way' and the light waves emitted have a consistent phase relation which does not change with time, Figure 12.5, shows coherent waves cited by, [5]. This is called 'temporal coherence' and is the most important characteristic of laser light.

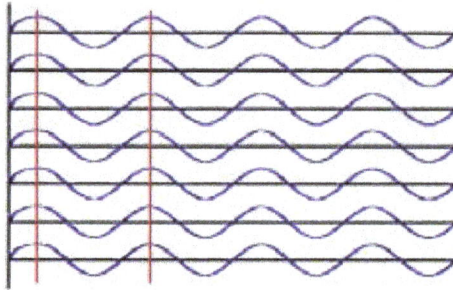

**Figure 15.5: Coherent Waves, cited by [5].**

Lack of coherence makes ordinary light an 'optical noise'. But coherence makes a laser light ' optical music '. Because of this coherence, tremendous amount of power of the order of $10^{13}$ Watts can be concentrated in a narrow space of linear dimension of $10^{-6}$m.

## 3. Directionality

The light from a laser is highly directional, which means that the light emitted is very tight, concentrated, and intense. In contrast, the light from a flashlight or a light bulb, for example, is comparatively diffuse and weak, since the light emitted is scattered in many directions.

The conventional sources like lamp, torch light and sodium lamp emit light in all directions. This is called 'divergence'. Laser, on the other hand, emits light only in one direction. This is called 'directionality' of laser light.

An example is the powerful search or guide light. If the beam from it travels a distance of 1km, it spreads to about a kilometer in diameter. If a laser travels a distance of 1km, it spreads to a diameter less than 1 cm. The directionality of laser enables us to focus the light to a point on a target at large distance.

## 4. High Intensity

The intensity of a wave is the energy per unit time flowing through a unit normal area. The light from an ordinary light source spreads out uniformly in all directions and forms spherical wave fronts around it. If you look at a 100 watt lamp filament from a distance of 30cm, the power entering your eye is less than 1/1000 of a watt. In the case of a laser light, energy is emanated in small region of space and in a small wavelength range and hence is said to be of great intensity.

Looking at a beam of a laser directly, allows all the power in the laser to enter the eye. Thus, even a 1 watt laser would appear many thousand times more intense than 100 watt ordinary lamp. For certain lasers, the intensity is so enormous that a power of $10^{15}$ watt can be concentrated into an area of 1 square centimeter [5 and 6],. Figure 12.6, illustrates the variations between laser and conventional lights, cited by [7].

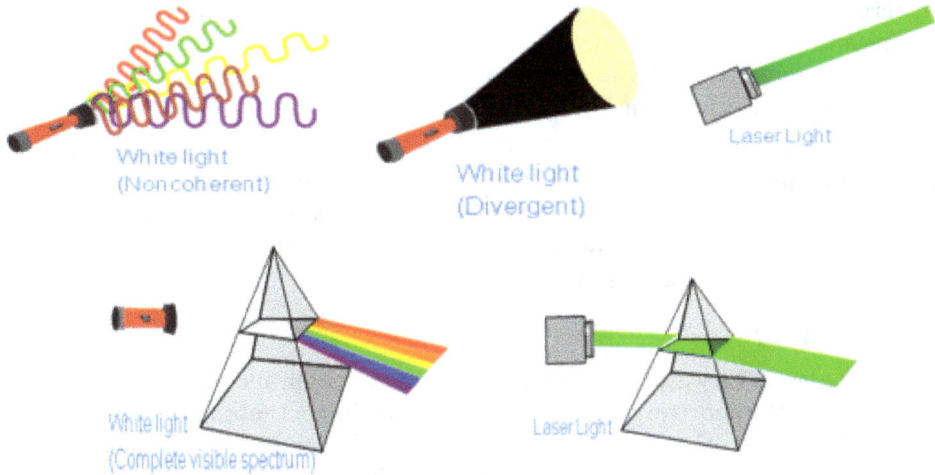

**Figure 12.6: Laser Beam Characteristics**
**(Variations between laser and conventional lights. cited by, [7].**

## Types of Lasers

According to the active medium, lasers broadly divided into four categories-solid lasers, gas lasers, liquid lasers and semiconductor lasers [8].

1. Solid-state laser is one in which the active centers are fixed in a crystal or glassy material.Solid state lasers are electrically non-conducting. They also called doped insulator lasers to avoid connotation of semiconductor.

2. Gas Lasers: Gas lasers are the most widely used lasers and the most varied. They range from the low power Helium-Neon (He-Ne) laser used in college's laboratories to very high power carbon dioxide laser used in industrial applications.

3. Liquid lasers or dye lasers belong to the family of liquid lasers. The active material is a dye dissolved in a host medium of a liquid solvent, such as ethylene glycol.

4. Semiconductor diode lasers can transform electric energy to light using the same principle as a *light-emitting diode LED*, but with internal reflection

capability, thus forming a resonator where a stimulated light can reflect back and forth, allowing only a certain wavelength to be emitted. Because a semiconductor diode laser has low energy density, this laser is often used for low-level laser therapy, without imparting any thermal effect [9].

A semiconductor laser consists of a flat junction of two pieces of semiconductor materials, each of which has been treated with different type of impurity. The p-type in the semiconductor materials has a deficiency of negatively charged free electrons in the crystal structure. This deficiency exists in the form of positions in the crystal that can accept an electron if one is available. These positively-charged holes are the carriers of electric current in p-type semiconductors. While the n-type semiconductor materials have a surplus of electrons that act as current carriers. If two slabs, one of p-type and one of n-type semiconductor material, are joined together, the result is called p - n junction, when an electrical current is passed through such a devices, laser light emerges from the junction region [2].

Lasers can also be classified according to the wavelength of the emitted radiation referring to infrared lasers, visible Lasers, UV and X- ray Lasers [5], or depending on their power so the lasers can be of high power or low power or what are well known as low level, cold or low intensity light therapy (LILT), phototherapy, light therapy, low-energy laser therapy, photobiomodulation [10].

## Laser Safety

While lasers vary greatly in power output, wavelength and purpose, the hazard potential for eyes and skin can be significant due to the concentrated energy density. *AS2211:2004 Safety of Lasers Products* is the principal document for laser safety. Lasers are divided into seven classes according to accessible emission limits. Modifications can increase the class and subsequent hazard of a laser.

*Class 1 lasers* are safe under most circumstances and are incapable of damaging the eyes or skin because of either engineered design or inherently low power output.

*Class 1M* lasers emit in the wavelength range 302.5 - 4000 nm and may be hazardous if optics are used in the beam.

*Class 2* lasers emit in the visible wavelength range 400 – 700 nm and have sufficient power output to cause damage to the eyes if viewed continuously. However, their outputs are low enough where eye protection is afforded by the blinking reflex. Additional hazard control measures take the form of cautionary signs or labels.

*Class 2M* lasers are similar to Class 2 however viewing may be more hazardous if the user employs optics within the beam.

*Class 3R* lasers emit in the wavelength range 302.5 – 106 nm and have the potential to cause damage to the eyes from intra-beam viewing but the risk is lower than for Class 3B lasers. Precautions are required to prevent both direct viewing and viewing with optical instruments.

*Class 3B* lasers are more hazardous because of either higher output or operation outside visible wavelengths. In addition, specular reflections *i.e.* non-diffuse surface reflections may also be hazardous. In general, more stringent controls are needed to prevent exposure.

*Class 4* lasers are high power devices capable of producing eye damage even from diffuse reflection. They may cause skin injuries and could also constitute a fire hazard [11], Figure 12.7 shows the warning symbol of lasers.

**Figure 12.7: Warning Symbol of Lasers.**

*Maximum Permissible Exposure MPE* limits are those levels of laser radiation to which, in normal circumstances, persons may be exposed without suffering adverse effects. A guide to MPE levels are given in *BS EN 60825-1*. They are based upon biological data collected to date, However, the risk assessment must also consider other hazards associated with use of the equipment, the main ones being:

1. Electrical - high voltages
2. Chemical/Fume - sample preparations, use of dyes and solvents, fumes produced by cutting and engraving.
3. Mechanical - manual handling materials and gas cylinders, falls when accessing high parts of the equipment, slips and trips on cables, cryogenic liquids, noise, vibration
4. Fire - high power, direct and reflected beams, ignition of cut materials
5. X-rays/electromagnetic interference.

All stages of the installation use routine maintenance, periodic servicing, and disposal of the equipment must also be considered in the risk assessment, whether this is carried out by staff, students or contractors [12].

## Laser Applications in Life

Lasers have many important applications. They are used in medicine and surgery. They are used in common consumer devices such as DVD players, laser printers, and barcode scanners and in industry for cutting and welding materials. They are used in military and law enforcement devices for marking targets and measuring range and speed. Laser lighting displays use laser light as an entertainment medium. Lasers also have many important applications in scientific research. In the following paragraphs, some of the most important applications of laser highlighted.

## Lasers in Medicine

### Laser Tissue Interaction

When laser light strikes a tissue surface, it can be reflected, refracted, scattered, absorbed or transmitted. The fractional intensity that goes into these different processes depends on the optical properties of the tissue like it's reflectivity, scattering and absorption coefficients, particle size, optical homogeneity, as well as the laser parameters like wavelength, energy, pulse duration, operation mode and output spectral profile.

Reflection is defined as the returning of the electromagnetic radiation by surface upon on which it is incident. While refraction occurs, when the reflecting surface separates two different indices of refraction. It originates from a change in speed of light waves.

In medical laser applications, refraction plays a significant role only when irradiating transparent media like corneal tissue. In opaque media, usually the effect of refraction is difficult to measure due to absorption and scattering [2].

Absorption, the amount of energy that is absorbed by the tissue depends on the tissue characteristics, such as pigmentation and water content, and on the laser wavelength and emission mode. In general, the shorter wavelengths (from about 500-1000 nm) are readily absorbed in pigmented tissue and blood elements [13].

Scattering can be defined as a change in direction of a light ray without a change in wavelength [14], scattering occurs during cold laser treatment and is considered to be a change in light propagation direction and thought to occur due to the varying shapes of biomolecules and varying tissue interface configurations [15].

Transmission is defined as the passage of radiant energy (light) through the tissue without any attenuation [16], Figure 12.8, shows the laser tissue interaction.

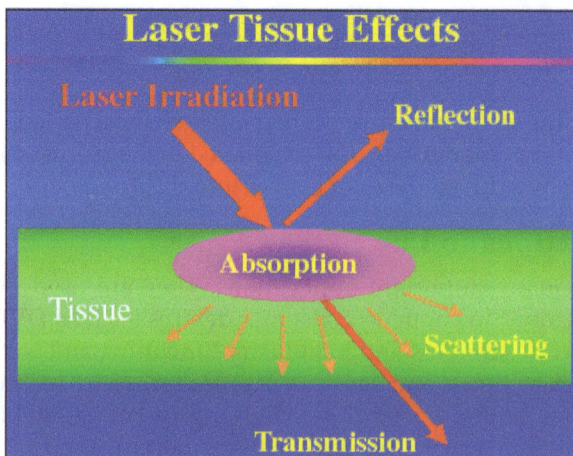

**Figure 12.8: Laser Tissue Interaction, cited by [13].**

Various types of lasers are used in medical diagnosis, treatment, or therapy. Types of lasers used in medicine include in principle any laser design, but especially, $CO_2$ lasers, diode lasers, dye lasers, excimer lasers, fiber lasers, gas lasers and free electron lasers. The use of lasers in medical procedures can be grouped into two distinct categories: diagnostic or imaging applications and therapeutic [17]. Nowadays lasers have been adapted to many medical procedures ranging from diagnosis, to medical treatment reaching to surgery interventions, oncology, physiotherapy, dentistry, dermatology and biostimulation [18].

Laser surgery is a highly sterile process since contact does not occur between the surgical tools and the tissues being cut. Further advantage is that the laser does not only cuts but also "welds" blood vessels being cut. Operations are done very fast and patients do not feel pain.

The first big success of lasers in medicine was in the treatment of eye. Argon laser has been in use for several years to treat the detachment of the retina. Laser photocoagulations of retinopathies are a common treatment. In people suffering from diabetes, for some unknown reasons, abnormal blood vessels spread across the surface of the retina. These blood vessels are very fragile and leave blood into the clear liquid of the eye. This process causes gradual dimming of vision. If these blood streamers are not removed, the patient can become blind. Laser photocoagulation is used to destroy areas of new blood vessels and areas of hypoxic retina which is believed to further increase blood vessels proliferation.

In dermatology, laser is used to cause homeostasis, the cessation of bleeding and removal of warts, freckles, acne and various other growths both malignant and benign, one typical example is bleaching birthmarks.

Lasers are used in destroying kidney stones and gallstones, and selectively destroy cancer. In the treatment, a dye called hemotopophyrin derivative is injected into the patient's body, the healthy cells flush out the dye while the dye is concentrated in cancer cells, which then scanned with laser irradiation causing fluorescing of the injected dye and changing of the $O_2$ into $O_3$ which is toxic and cause damage of the cancer cells [8], medical areas that employ lasers include:

☆ Angioplasty

☆ Laser Acupuncture, Figure 12.9.

☆ Cancer diagnosis

☆ Cancer treatment

☆ Cosmetic applications such as laser hair removal, tattoo removal and laser liposuction

☆ Dermatology

☆ Dentistry, Figure 12.10.

☆ Lithotripsy

☆ Mammography

☆ Medical imaging

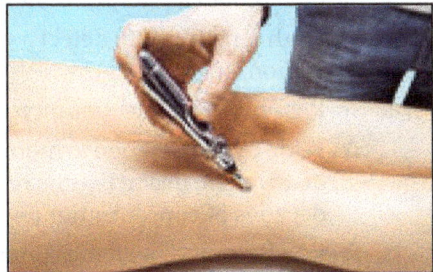

**Figure 12.9: Laser Acupuncture.**

☆ Microscopic surgery

☆ Ophthalmology (includes Lasik and laser photocoagulation), Figure 12.11.

☆ Optical coherence tomography

☆ Prostatectomy

☆ Endovenous laser therapy

☆ Laser-assisted new attachment procedure

☆ Laser interstitial thermal therapy

☆ Light therapy

☆ Low level laser therapy

☆ Photodynamic therapy

☆ Photomedicine

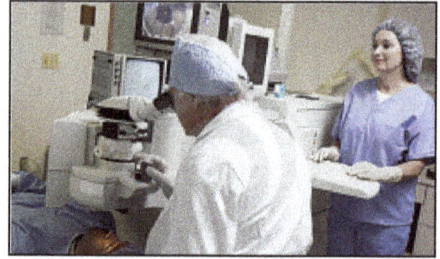

**Figure 12.10: Vision Errors Correction by Laser.**

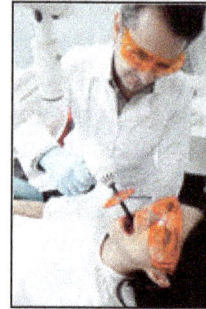

**Figure 12.11: Laser Used in Dentistry.**

## Lasers and Pollution

Lasers have revolutionized spectroscopy, greatly expanding the field of laser spectroscopy in many areas. Immediately after the discovery of the laser in 1960, the laser systems were developed for atmospheric studies. The most common principle employed for the detection of environmental pollutants involves the interaction of the trace species with laser light (absorption, scattering). Up to now, various diagnostic methods based on physical process caused by laser light–environmental species are developed.

Ideally, detection and monitoring laser based spectroscopic technique should fulfill the following requirements:

☆ High selectivity, for a particular gas species, with no observable cross-response from other species; to measure accurately trace gas concentrations of less than a part per billion.

☆ High sensitivity, to detect a very low concentration, below ppt (parts-per-trillion, 10-12).

☆ Possibility to detect numerous compounds with one instrument.

☆ Wide dynamic range to monitor high and low concentrations with a single instrument, in a real–time response that can be linear over more than four decades of concentration.

☆ Fast Response, with measurement speeds of fractions of a second, or signal averaging to achieve still higher sensitivity.

☆ Good temporal resolution for on–line monitoring.

Remote sensing technique, use of lasers to measure samples at some distance from the laser system. The advantage is the contactless measurement, measurements results give three – dimensional concentration or integrated profile of pollutions.

An important technical criterion of spectroscopic method is the wavelength region used: visible, ultraviolet (UV), infrared (IR) and microwave spectroscopy. For species identification IR region is particularly attractive, because the most of main atmospheric molecules ($H_2O$, $CO_2$, $O_3$, $CH_4$, $N_2O$, CFCs), have well defined and highly characteristic spectral features in the mid-IR spectral region (the so called fingerprint region) where molecular line intensities are reasonably large [19].

## Iraq and Scientific Movement

Iraq lies in the Middle East spanning 437,072 km$^2$ and total population of 31,234,000 on 2009, Figure 12.12. It is a member of the Non-Aligned Movement (NAM) and was one of those which struggle hardly to initiate it, yet it is one of the most active members in this movement and it's affiliate foundations, organizations and centers specifically the Centre for Science and Technology of the Non-Aligned and Other Developing Countries (NAM S&T Centre), Figure 12.13, shows the distribution of NAM countries in the world.

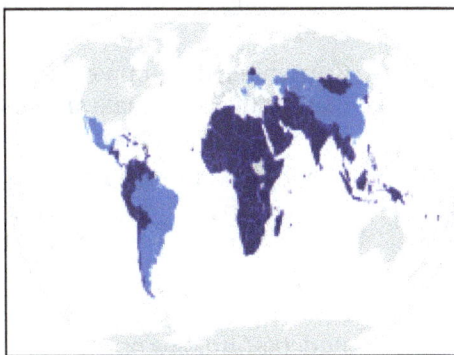

**Figure 12.12: Iraq Lies in the Middle East.   Figure 12.13 : NAM Countries in the World.**

Many events went on in my country since the eightieth of the last century, beginning with the Iraq- Iran war upward to removing of the regime in 2003.

Unfortunately these events alter the growing and development of all the life's fields especially the scientific research, they deteriorate the sustainable development and the infrastructure of the scientific process of a country which was once the capital of the world and the lodestar for science, culture and education.

In our country we have an Institute of laser science belongs to Baghdad University and a Faculty of Laser and optical belongs to University of Technology, we also initiated a society specialized in laser applications called Iraqi Laser Society.

## Conclusions

There is a deep sense of need to keep up with modern science and evolution knowledges, particularly those that make us taking the right path toward the progress and development with confidence, like laser and nanotechnology.

# Recommendations

## Specialized Center for Scientific Research and Treatment with Laser

There is a need for more efforts and hard works to reach the science margins in this field, we need cooperation with all the neighbors' countries to initiate a large specialized Center for Scientific Research and Treatment with Laser in our country undertaking the following tasks:

1. Dissemination of scientific knowledge in the field of lasers through framing scientific activities in the areas of research and studies, seminars, scientific conferences and science fairs.

2. Information documentation and exchange of experience among workers in the field of laser from all disciplines (medical, engineering and physical) inside the country and with the neighbors and other NAM countries.

3. Scientific relations with corresponding centers regionally, within the organization and globally and everything would take care of and develop competence.

4. Curriculum development for the preparing intermediate stages enriching them with appropriate amounts of informations regarding laser through adding and the new and developing present.

5. Moving toward the media (newspapers and journals), television and radio to customize programs for the definition of the community this type of energy which we expect him to enter into the details of daily life of the individual.

6. An ambitious plan to initiate a medical laser institute or college grants the students (diploma or Bachelor of medical lasers in all branches of medicine known). With completion of this project then we shall move towards the establishment of a specialized hospital including all the diagnostic, therapeutic and surgical specialties which may serve all the Middle East countries.

7. Fields of research include:

   ☆ Environmental and pollution researches.

   ☆ Communications research.

   ☆ Space research.

   ☆ Dimensions and measurements research.

   ☆ Researches regarding to increase the agricultural and livestock production and combat pests.

   ☆ Preparation of vaccines for diseases that strike the zone (the country and it's neighbors).

   ☆ Genetics and breed improving and treatment of infertility researches.

   ☆ Laser uses in art and city beautification and illumination.

   ☆ Laser in industry specially the electronics and optical.

   ☆ Research involving interventions of lasers in biology.

☆ Research concerning the climate, weather prophesying BA earthquakes and volcanoes.

☆ Crime detection, fingerprints and fumbling with explosives.

In the field of medicine laser, it is useful to say that our Government has brought a lot of medical devices for treating Visual errors, lithotripters, surgical and gynecological laparoscopic systems but yet these devices remain just little compared to the need, and usually the physicians runs these devices did not receive adequate training, so we need cooperation with neighboring countries in the framework.

## REFERENCES

1. John, C. Fisher, 2007 ; Fundamentals of Laser Physics, Optics, and Operating Characteristics for the Clinician, Part One ; Basic Science and Safety - Chapter 1, Pp.: 3- 21, The American Board of Laser Surgery Inc., 2007.

2. **Issa,** M. M., 2005 ; The Evolution of Laser Therapy in the Treatment of Benign Prostatic Hyperplasia. *Med. Reviews, LLC.*7(9) : 15 – 22.Cited by Zainab, A.R., 2012; Preparation of Vaccine for Diabetic Foot Pathogenic Bacteria using Low Level Diode Laser, M.Sc. Thesis, College of Science – Al – Muthanna University.

3. Zungu, L.I., Abrahamse, H. and Evans, D.H.,2008; Mitochondrial responses of normal and injured and human skin fibroblasts following Low Level Laser Irradiation, Ms.C. Thesis. Faculty of Health Science, University of Johannesburg. Cited by Dunnia, A.B., 2011; Effect of 820nm Diode Laser on Some Hormones and Enzyme Concerning with Wound Healing and Skin Loss Sealing, M.Sc. Thesis, College of Science – Al – Muthanna University.

4. Melinda Rose, 2010; A History of the Laser: A Trip Through The Light Fantastic, Photonics Spectra, Senior Editor, melinda.rose@photonics.com. http//www. photonics.com.

5. Svelto, O., Di Milano, P., Fisica, D. and Davinici, P.L., 2010; Principles of lasers, Introductory Concepts, Properties of laser beam, Chap; 1, Pp.: 8-13. 5th.ed., ISBN;978-1-4419-1301-2. Cited by Dunnia, A.B., 2011; Effect of 820nm Diode Laser on Some Hormones and Enzyme Concerning with Wound Healing and Skin Loss Sealing, M.Sc. Thesis, College of Science – Al – Muthanna University.

6. Ganesh, K., 2010; Properties of Laser Light, Introduction to properties of laser light, Tutor.Vista.com., Management Team Bios. http//:www. TutorVista. com./questions. Cited by Dunnia, A.B., 2011; Effect of 820nm Diode Laser on Some Hormones and Enzyme Concerning with Wound Healing and Skin Loss Sealing, M.Sc. Thesis, College of Science – Al – Muthanna University.

7. Martinm, M., 2008; Lasers and Light Surgical Marketing Manager Lumenis Europe. German Conference held in 2008, mmartin@lumenis.com. Cited by Dunnia, A.B., 2011; Effect of 820nm Diode Laser on Some Hormones and Enzyme Concerning with Wound Healing and Skin Loss Sealing, M.Sc. Thesis, College of Science – Al – Muthanna University.

8. Avadhanulu,M.N., 2009; An Introduction to lasers Theory and Applications. Published by S. Chand and Company Ltd,.7361, Ram Nagar, New Delhi-110055. ISBN:81-219-2071-X,An ISO 9001:2000Company. Pp.:166-174.

9. Choi, E. J., Yim, J. Y., Koo, K. T., Seol, Y. J., Lee, M. Y., Rhyu, I. C., Chung, C. P. and Kim, T., 2010; Biological effects of a semiconductor diode laser on human periodontal ligament fibroblasts. *Journal of Periodontal and Implant Science.* 40 (4):105–110.

10. Dais, J., 2009 ; Low Level Laser Therapy Position Paper for the CMTBC: An Examination of the Safety, Effectiveness and Usage of Low Level Laser Therapy for the Treatment of Musculoskeletal Conditions, LLLT Position Paper.

11. Occupational Health and Safety Information Sheet, OHS, 2009; Laser classification.

12. Laser Safety, 2009; University of Greenwich, safetyunit@gre.ac.uk.

13. Jyoti, N., Pankaj, M., Tulika, G. and Shelly, A., 2010; Dental Laser – A boon to prosthodontics. International Journal of dental clinics 2 (2): 13–21. Cited by Wafa'a Abdulmutalib Naji, 2014; Effect of 915nm diode laser on some hormones and minerals concerning with fracture healing, M.Sc. Thesis, College of Science – Al- Muthanna University.

14. Fisher, J.C. 2007. Qualitative and quantitative tissular effects of light from important surgical lasers: optimal surgical principles, Interaction of laser light with living tissue, Chap. 4, PP: 58-81, The American Board of Laser Surgery Inc.

15. Tiziano, M. 2004. Cold lasers in pain management. Practical Pain Management:1-5.144

16. Markolf, H. N., 2007; Laser-tissue interactions: fundamentals and applications, 3rd revised edition. Ch.2-3 Pp.: 9-100. Cited by Wafa'a Abdulmutalib Naji, 2014; Effect of 915nm diode laser on some hormones and minerals concerning with fracture healing, M.Sc. Thesis, College of Science – Al- Muthanna University.

17. Wells, J., Kao, C., Konrad, P., Tom, M., Kim, J., Jansen, A., and Jansen, D., 2007; Biophysical Mechanisms of Transient Optical Stimulation of Peripheral Nerve. *Biophysical Journal.* 93 (2) : 2567–2580. Cited by Zainab, A.R., 2012; Preparation of Vaccine for Diabetic Foot Pathogenic Bacteria using Low Level Diode Laser, M.Sc. Thesis, College of Science – Al – Muthanna University.

18. Valchinov, E. S. and Pallikarakis, N. E., 2005 ; Design and testing of low intensity laser biostimulator. *Bio-Medical Engineering OnLine* 4(5). Cited by Zainab, A.R., 2012; Preparation of Vaccine for Diabetic Foot Pathogenic Bacteria using Low Level Diode Laser, M.Sc. Thesis, College of Science – Al – Muthanna University.

19. Mira Terziæ, Janez Mo ina and Darja Horvat,2006; Using laser to measure pollution. Facta Universitatis, Series: Physics, Chemistry and Technology Vol. 4, No 1, 2006, pp. 71 - 81

# — *Section IV* —
# Theoretical Frameworks

## Chapter 13

# A Hybrid Model for Integrating S&T Policy with S&T Diplomacy

*Tahereh Miremadi*

*MAPSED Research Center for S&T Policy and Diplomacy*
*Iranian Research Organization for Science and Technology (IROST), Iran*
*e-mail: miremadi@IROST.org; tamiremadi@yahoo.com*

## ABSTRACT

This paper aims to build a theoretical model to explain the interactive nature of dynamism of domestic public policy and diplomacy in the domain of science and technology. Bridging two different frameworks of Advocacy Coalition Framework and Double Edge Diplomacy, the paper attempts to show how domestic controversial policy advocacy of S&T determine the alternation of a country's position at the international arena and how the factor of policy brokering stabilizes this position by solidifying the relation between domestic and international policy communities. The presented model is applied on the case study of nuclear energy in Iran.

*Keywords:* *Iran, Advocacy coalition framework, Double edged diplomacy, Nuclear science and technology policy and diplomacy.*

## 1. Introduction

The long-awaited, landmark Geneva interim accord reached last November between Iran and the P5+1 powers was, to some extent, the culmination of years of conventional diplomacy coupled with the systematic pressured imposed on Iran. However, it must now be abundantly clear that the agreement would not have materialized without Iranian government's bold decision to adjust the central theme of its foreign policy, seeking to reconcile its strategic national interest with that of the international community. What are the causes of this policy changes? Is it a temporary or a permanent paradigm shifted in the S&T diplomacy of the country?

This paper is devoted to analyze the internal relationship between the Iranian diplomatic contingent involved in the negotiation process and the domestic political forces. The aim of the paper is twofold First by probing the underlying social dynamism of nuclear technology's policy subsystem and its interaction with other subsystems, the paper addresses two deeply rooted co-joint learning processes which ultimately alter subsystem policy settings. One, learning process resulting from increasing major reflexive and institutional capacity of Iranian political elites at the level of society, and on the other hand, at the level of policy analysis, which has been brought forth by strategic interaction of people within the policy research and policy making dimensions. The latter was partially the result of formal policy analysis and trial and error learning amid international pressure as well as overall mismanagement of the economy. The first part of the paper concludes by showing how this interim accord is an internationally outward manifestation of a domestic long term cumulative learning in the field of the nuclear S&T technology policy in Iran and not just a swing of the policy pendulum –fueled by elite dissensus over means and ends of Iran's diplomacy pursuant to the 11th Presidential election of Islamic Republic of Iran in 2013.

The second objective is to explain how the learning factor has positively affected Iranian bargaining power and ultimately augmented the plausibility of an accord between Iran and the West. This objective may be viewed as an extension to what Evans and Putnam have called "domestic win–set" (Putnam, 1993)and hence decreasing the probability of defect on either side.

The paper continues with introducing our theoretical framework, which is built by integrating two different theoretical constructs: Double Edged Diplomacy (hereafter DED), and Advocacy Coalition Framework (hereafter, ACF). In this manner, the paper builds a two-dimensional model of science and technology oriented public policy and foreign policy. Based on this hybrid model, we organize our data concerning the subsystem of nuclear technology policy in the first section, and in the next section, the acceptability set of Iranian negotiating party culminated in the interim accord is analyzed and explained.

## 2. Reviewing the Literature

There are different theoretical frameworks within the public policy discipline which explain policy learning process. (Borass, 2011), (Hall,1993)and Howlett and Ramesh (Howlett, 2002).

Among them, Paul Sabatier's Advocacy Coalition Framework (ACF), has been recognized as distinctive framework because of its attention to social dynamism by taking public policy as the reflector of the belief systems of the advocacy coalition within a policy subsystem. It has been developed to explain policy change and continuity through internal dynamism and external factors.

A major assumption of ACF is that, actors in a policy domain or subsystem can be aggregated into a few advocacy coalitions. These coalitions typically consist of interest group leaders, agency officials, legislators, applied researchers, journalists, and politicians. Parties within a coalition share a set of normative and causal beliefs and show a non-trivial degree of coordinated behavior to realize their objectives and policy proposals (Sabatier, 1998).According to Sabatier, belief systems of these coalitions are organized in a hierarchical, tripartite structure. The deep core of a belief system includes basic ontological and normative beliefs. The policy core represents basic normative commitments and causal perceptions across an entire policy domain. These beliefs concern the basic perceptions of the seriousness of a policy problem, its main causes, and perceptions about the appropriateness of institutional arrangements to deal with this problem. Finally, the secondary aspects of a belief system are the less than subsystem-wide beliefs concerning problems, causes, and remedies.

ACF presents several specific hypotheses about conditions that are conducive to cross-coalition policy oriented learning. It argues that analytically tractable issues, an intermediate level of informed conflict, and the presence of professional forums prestigious enough for members of opposing coalitions to participate in, are conducive to learning (Jenkins-Smith and Sabatier, 1993).[1] According to Sabatier, information about the nature and complexity of the problem is essential to inform policy decision-making. Therefore, administrative agencies, legislators, and analysts need to know and understand as much about these problems, its causes, and the likely impacts of various interventions as they deliberate, craft, and implement public policies. This suggests a role for technical specialists in policy activities. Second, Advocacy Coalition Framework argues that understanding policy change and policy learning requires a decade or more. This premise is based on Weiss' research that enucleate the importance of the "enlightenment function" of policy research showing how learning over time can alter the perceptions of policymakers. (Weiss, 1999)

And finally, an important feature of the ACF is "policy brokers" who do not lean towards any competing advocacy coalitions, but seek compromise between them. They are generally conceived of elected officials, senior civil servants, and regulatory bodies.

As much as ACF is an ideal basis for analyzing the domestic side of new development in the Iranian nuclear policy, it cannot stretch itself to cover the international side since this framework is generally applied within a disciplinary context that views policy formation as an essentially domestic level process occurring within states. According to ACF policy shifts are the result of changes external to the policy system, including dynamic system events at the international level. (Sabatier and Jenkins Smith 1993).

That is why we have to search for a complementary basis to frame the international dimension of our study. It suffices to note that the existing literature

---

1  Across-coalition learning is generally easier in the natural than in the social/behavioral sciences because the theories and accepted methods are better established and the objects of study are not themselves actors in the policy debate. According Sabatier and Zafonte (2001). The existence of a forum that is (a) prestigious enough to force professionals from different coalitions to participate and (b) dominated by scientific norms: The latter assures a general consensus on the appropriate rules of evidence, as well as some attention to underlying assumptions.

linking the domestic and international narratives level has evolved through three periods:

In 1969, was the idea of "convergence" presented by James Rosenau (Rosenau 1969) focusing on the overlap between domestic and foreign affairs as a result of ICT revolution and globalization. He addressed the blurring distinction between some domestic and the international arena and called this phenomenon a "Conversion".

In 1970s and 1980s, "second image" and "second image reversed" literatures explored the domestic causes of foreign policy and the international sources of domestic policy respectively. (Gourevitch, 1978), (Katzenstein, 1983)).

In 1990s, the two level games approach proposed an interactive model by viewing national negotiators as constrained simultaneously by domestic/foreign divide. (Evans, 1993)

The paper is hardly able to follow the lead of Rosenau's "convergence" theory. Contrary to "Convergence" phenomenon, and providing the fact that "nuclear technology" for many Iranians are a matter of national pride and serves an important purpose in their national security and the defense of their country's sovereignty against foreign intrusion, we are in fact, dealing with the phenomena of "divergence" in the relationship of the public policy of Iran and the outside world.

Despite the fact that the second set of theories looks very promising, since they neither could deal with the negotiation capacity of the statesmen in the negotiation, nor could be good choice for this paper

Therefore, we have chosen Peter Evans (Evans, 1993) approach of double edge diplomacy to benefit from his clear-cut, yet interrelating divide between the national and international affairs. Moreover, he discusses different situations in which the international negotiation can have different result.

In consideration of the theoretical construct of the two level games (hereafter DED) approach of Evans, Jacobson and Putnam (Evens Peter, 1993), we find that there are three essential building blocks: specifications of domestic politics or the nature of win-set, the international negotiating environment, and the statesmen's preferences, (Evens Peter, 1993):23. According to this approach, the executives have 'Janus face' dealing with both 'constituency' driven domestic and the international system logics. (Evens *et al.*, 1993):5

The authors assume that diplomatic strategies and tactics are constrained both by what other states will accept and by what domestic constituencies will ratify. On the other hand, the diplomats in their attempt try to build an international agreement, seek simultaneously to manipulate domestic and international politics. (Evans *et al.*, 1993): 5. Defining diplomacy as a process of strategic interaction, the authors contend that actors simultaneously use these actions to take account of and, if possible, influence the expected reactions of other actors, both at home and abroad.

It must be emphasized that there are substantial differences between DED and ACF. Among these differences are:

1. The DED assumes the formation of coalition and interest groups are based on an assessment of the relative costs and benefits of negotiated

alternatives to the status quo (Evens Peter, 1993):24, while our model inspired by AFC presupposes people engage in politics to translate their beliefs into action.

2. DED contends that the basis of these cost and benefits assessments forming the interest groups remains constant throughout the analysis (Evens *et al.*, 1993): 24, while our model based on ACF describes how policy can change. It presumes that the stakeholders' beliefs have three layers with different capacity to change

3. ACF underscores the importance of learning and enlightenment factors in the policy change. In fact, learning factor is the first of 5 premises of the initial version of the ACF. Moreover, technical information concerning the magnitude and facet of the problem, its causes, the probable impacts of various solutions, are assumed to play an important role in many administrative agencies. (Sabatier and Jenkins-Smith, 1990)

4. According to the DED model, the set of agreements preferred by statesmen to the status quo may be termed the statesmen "acceptability-set"(Evans *et al.*, 1993): 30.The focus of the analysis is on the strategic incentives created by certain configurations of the acceptability set relative to the domestic win–set. The possible configurations can be divided into three categories: the statesmen as agents, as doves, and as hawks. In the case of statesman as dove, the acceptability–set lies at least partially outside the domestic win-set and closer to the opposing win-set. In the case of the statesman as hawk, the acceptability–set lies at least partially outside the domestic win-set but further from the opposing win-set than the set of rectifiable agreement.

Our presumption, inspired by ACF, is that the statesmen acceptability-set reflects the core beliefs of the advocacy coalition of the chief of the government (COG) or the advocacy coalition which runs the executive branch. Depending on the settings of the political system, the domain that the COG can influence is different from the domain controlled by other major players of the State like the Parliament, the Courts, or Supreme Leader, all of which are required for rectification of international deals, and if COG and other players belong to two competing coalitions, their difference of acceptability- sets makes the size of national win-set small and its ratification unpredictable. When there is a devil shift between two rival coalitions, the size of acceptability-set of the statesmen is dependent on the political, informational and economical resources of the advocacy that runs the executive branch and his political and policy rival. If they are countervailing powers, the agreements between the negotiating team and its foreign counterpart is at risk of domestic refusal. Contrarily, the size of the win-set is the largest if the COG is not a policy advocate, but policy broker who helps to mediate between two core policies.

Coupling ACF with DED, we build a theoretical model to determine how achange of public policy can modify the behavior of diplomats and alter their tactics and strategies from zero-sum game to positive-sum game and vice and versa,and ultimately augment the plausibility of an accord between two negotiating parties.

## 3. Main Body

### 3.1 Defining Policy Problems

The commencement of Iran's nuclear activities back in the 1950s carried the full blessing of the Western countries. At that era, Iran was a junior partner of the United States as a member of the Central Treaty Organization (CENTO) and had a close relationship with the US as well as other Western countries. In 1957, in the spirit of President Dwight D Eisenhower's program of Atoms for Peace, which aimed to spread the peaceful use of nuclear technology in the world, the US government was instrumental in forming the International Atomic Energy Agency under the auspice of the UN. Iran signed the UN Nuclear Non-Proliferation Treaty in 1968 on the day it was opened for signature (Nikou, N/A). The cooperation started with the United States decision to supply Iran a research reactor for medical uses based on the Atoms for Peace program. In the same year, the Institute of Nuclear Sciences, affiliated with the Central Treaty Organization (CENTO), was relocated from Baghdad to Tehran University. The Institute became a training center for Iranian students as well as those from Pakistan and Turkey (Entessar, 2009). The government of Iran signed a contract with the American Machines and Foundry Company (AMF), for the construction of a research nuclear reactor at the University of Tehran (6 megawatts) in 1960. Accordingly, the first center for nuclear energy in Iran, the Tehran Nuclear Research Center, was established at the University of Tehran in the same year. Later in 1972, based on the recommendation of the American government, the Atomic Energy Organization of Iran (AEOI) was established. At the same year, the university of Tehran and University of Shiraz became active in teaching nuclear technology and some Iranian students were sent abroad by the government to study in the field of nuclear energy. In 1975, the Massachusetts Institute of Technology signed a contract with the AEOI to provide training for the first cadre of Iranian nuclear engineers and scientists.

Based on the recommendation of the Stanford Research Institute (SRI), on Nov. 1974, Iran signed an agreement with Kraftwerk Union AG (KWU), a subsidiary of Siemens, to construct two light water reactors with the capacity of 1300 Megawatts. Construction began the next year, and completion was scheduled for 1981 (Sich, 2012). More than two thousands German and Iranian experts began working on this project together, which at the time was one of the largest nuclear energy production plants in the world.

French companies have also played an important role in the introduction of nuclear technology to Iran. France formally began its nuclear activities in Iran in 1977, and in October of that year, Iran and Fram Atome, a French company, forged an agreement for the construction of two nuclear power plants with a 900 megawatt capacity near Ahvaz. But before that in 1974, the Iranian government provided $1 billion credit/loan to the French Atomic Energy Commission to build the Eurodif plant. The loan would have entitled Iran to buy 10 percent of enriched uranium produced by Eurodif. In 1977, Iran paid an additional $180 million for future enrichment services by Eurodif, for the construction of the Eurodif factory, and the right to buy 10 per cent of the production of the site. (Olivier, 2006) Another example of the cooperation between Iran and France on nuclear issue was the Nuclear Technology Center at Esfahan (Isfahan) founded in the mid-1970s with French assistance in order to provide training for the personnel that would be working at Bushehr reactor site.

The establishment of the revolutionary Islamic government in 1979 ended U.S. participation in Iran's nuclear energy program. For its own part, the new government cut back or cancelled much of Shah's ambitious nuclear program including plans for power reactors. However, the political leadership's policy core system was changed during the war with Iraq, (1980-1988). The eight-year confrontation was the Middle East's bloodiest conflict in modern times.Iraq used chemical weapons against Iran and the West did not condemn Iraq nor did it provide Iran with the necessary self-defense capability (Chubin, 2006). Iran's nuclear program was an outgrowth of this experience.

In the Post-war era, Iran planned to be as self–reliant as possible in technology and the nuclear program was a part of that strategic shift. However, by the time Iran refocused on the nuclear program, the Western countries had changed their strategy about nuclear technology transfer to Iran for obvious reasons.

Iran made numerous attempts to acquire nuclear related technologies, parts, instruments, and materials. Many of these attempts failed to yield any successful results. Following disappointment with the West, Russia, China, and Pakistan emerged as the most important partners of Iran in providing spare parts, as well as human capital and training in the field of nuclear technology.

However, the redirection from the western to eastern countries for the Iranian procurement, if solved the Iranian problem, exacerbated the problem of Iranian nuclear program for the West. The main concern of the West can be classified into 3 main questions (Cohen, 2014, May 14):

☆ What size nuclear program is plausibly consistent with civilian use;

☆ How to ensure the barrier between such a program and militarization;

☆ How to achieve complete transparency and relentless verification.

The measures the West has taken to answer these questions and ease its concerns, required Iranian prompt respond.

## 3.2 Policies to Counter Measure the Western Policies

While analyzing the official discourse developed roughly about the nuclear issue in Iran, we generally find three sets of policies: The first set is the policies which are repeatedly announced and reiterated in every occasion when Iran gets the chance. It includes the advocacy of the peaceful use of nuclear energy which will indeed multiply the Iranian energy sources. According to the official documents(which official documents the name of those documents should be mentioned), the Islamic Republic of Iran is after nothing beyond its legitimate rights as stipulated in the NPT, wishing to enjoy its inalienable rights in return for meeting its obligations. The prime target of the Islamic Republic of Iran with respect to nuclear energy is the production of nuclear electricity. Due to the limited resources of fossil fuels, the right of future generations to use this energy resource and detrimental impact of use of fossil fuels on environment as well as growing population and economy and increasing need to energy resources, and preferable use of oil in processing industries, the Islamic Republic of Iran cannot remain dependent merely on the fossil fuels and has to diversify its required energy resources. To develop nuclear plants, the Islamic Republic of Iran has to produce 20000 megawatts electricity by 2025. The decision was made on the premises of growth factors of its economy as well as the Parliament's approval. Based on its long term plan, Iran has to provide the required fuel for its nuclear plants from internal and external sources.

Fortunately, there has been no international objection towards this goal The majority of the developing countries have supported the Iranian State's claim to peaceful nuclear technology. For Example, different statements issued by the Non Alignment Movement (NAM) in different meetings of the Board of Governors and the statement issued by the NAM Troika, as well as the statement issued by the NAM Heads of States and Ministerial Meetings along with the statement issued at the OIC meeting in Baku, all support of Iranian peaceful nuclear activities. (Times, 2014)

The second and third sets of policies, which have however difference with the first one, are related to some programs and technical questions. The latter two sets of policies can be considered as the means and the settings for attaining the previous sets. In our case, the policy to process the material like enriching uranium, or produce machineries and equipment, can be considered the policy instruments or policy means and the calibration of the instruments is the settings of the policy. The other disposition of these sets of policies is that contrarily with the previous one, there is no consensus upon which policy instruments should be applied and what are the policy settings. In fact, there is a remarkable disagreement at the domestic levels about which policy instruments and calibration should be employed to reach the developmental goal of indigenous nuclear energy. The detail of these questions is as follows:

1. Is it in Iran's interest to implement the additional NPT?
2. What size of nuclear program is plausibly consistent with the Iranian peaceful use? How many centrifuges? How many sites? How many kilograms of enriched uranium should the Iranians keep in their stockpiles?
3. How far should Iran going to be transparent in terms of nuclear and non-nuclear activities (missile development and its purpose.)

Because of these disagreements, two major advocacy coalitions have been formed. These two have different policy cores and each one have a distinct view about the policy instruments and its settings and calibration. Since each advocacy coalition has had the chance to win the presidential election, form the government and run the executive branch, each has had equal opportunity to translate its policy beliefs to technology policies. Consequently, Iran has witnessed different episodes in which different policies and plans, regarding the policy instruments, have been designed and implemented. In the next section, we review these episodes, in the history of Iran after revolution, with visible alternations at the second level of nuclear policies. As we can see, each episode has different degree of cohesion and consensus among political elites about the aforementioned policy instruments.

## Phase One: Period of Consensus 1987-1997

The period of maximum consensus on Iran's nuclear program which lasted for 15 years. The revival of the Shah's nuclear program was initially presented as necessary to the diversification of the country's energy resources. Nuclear technology was viewed as cutting edge development and indispensable for any self-respecting power. Throughout the program's early stages, there appeared to be a general consensus among the political elite about the need and the right to proceed. In a trial and error process, they tested every available venue to pave the way for technological capacity building of an independent nuclear technology, including the uranium enrichment. (Rouhani, 2011).

## Phase Two: Early Controversy 1997-2005

In this period, President Rafsanjani finished his term and President Khatami sworn as the fifth President of the Islamic Republic of Iran. The reformist government ameliorated the image of Iran in the international arena, in part with their tactics and strategies employed in nuclear diplomacy regarding the enrichment of uranium. Throughout this period, the nuclear program was largely concern of Iran's political elites. The Supreme National Security Council managed all aspects of policy regarding the program and its decisions, therefore, were said to be reflection of a national consensus. In this period, the reformist government of President Mohammad Khatami secured an agreement in the Supreme National Security Council to address international concerns and attain a compromise. Iran agreed to apply the NPT's Additional Protocol which permitted stricter international inspections and agreed to voluntarily suspend enrichment for a limited though unspecified period of time. But, the opposite advocacy coalition (conservatives) who gained control of Iran's parliament in 2004 began criticizing their political rivals (reformists) as being too soft towards the United States for compromising Iran's interests. In 2005, newly elected President, Mahmoud Ahmadinejad, revived the enrichment program, thus officially ending the previous administrations diplomatic overtures to the Europeans.

## Phase Three: Tough Stance 2005-2013

Iran's nuclear program became increasingly political during this phase. As of 2005, both the executive branch and the Parliament were dominated by a conservative coalition. Among hardliners, Ahmadinejad's rallies frequently included orchestrated chants in favor of Iran's nuclear rights. The President announced that Iran's nuclear program was "like a train without brakes," not vulnerable to outside pressure. However, two factors spurred intense backlash—and a reaction on the other side of the political spectrum. First, the United Nations imposed a series of U.N. resolutions between 2006 and 2010 that included punitive sanctions. Second, The United States and the European Union imposed even tougher unilateral sanctions. By 2010, the divide over Iran's nuclear program was principally about domestic political schisms rather than the desired strategies of key stakeholders.

**Phase Four: The Informed Consensus 2013**

Iran's 11th Presidential election was held in 2013. The results were unexpected. Iran's former secretary of the Supreme National Security Council and chief negotiator in 2003, HasanRouhani, was elected in a landslide victory. His mandate was clear: "tackling the issue of nuclear negotiation".

During his first 100 days of his Presidency, the Rouhani's administration changed the country's approach to the nuclear negotiation in both substance and tone, and the Foreign Ministry was chosen to implement the nuclear diplomacy by negotiating with P5+1.The evidence shows that nuclear policy and diplomacy of the President Rouhani is the policy of the middle man mediating between the long rival advocacy coalitions of reformists and conservatives. As a consequence, after more than 15 years, a new consensus has emerged in the political atmosphere. This consensus is different from the initial one in terms of the experience and analytical capacity used to near the positions of the two coalitions. That is why we can call it the 'informed consensuses, which means a consensus based on policy oriented learning and knowledge earned by trial and error and education, which culminated towards the relative alignment of the policy beliefs of the two advocacy coalitions in the nuclear policy subsystem.

## 3.3 The Reflection of the Consensus at the international Level

We now proceed to the international level and speculate that how this alignment reflected on the Iranian stance in the negotiation after the 11th Presidential election. Specifically, we should explain how the rapprochement between the two coalitions affected the configuration of the statesmen's acceptability–sets and the national win-set.

1.  The period between 1997-2004; in this period, Iran succeeded to make important progress in its indigenous nuclear efforts. By 2003, when the scope of its nuclear program became clear, Iran had already made progress towards mastering the technology needed to enrich uranium, one of the sensitive dual technologies. Facing scrutiny from the UN Security Council, Iranian officials and experts have argued that under Iran's safeguards agreement with the IAEA, it was only obligated to disclose nuclear activities six months prior to the introduction of nuclear material into a facility. This was a reference to Article 42 of Iran's Safeguards Agreement and a secondary document known as a 'Subsidiary Arrangement'. The language in the "Subsidiary Arrangement" required notice to the IAEA of new facilities "no later than 180 days before the introduction of nuclear material into the facility, and the provision of information on a new Location. On the other hand, the opposite party maintained that because many of Iran nuclear experiments were conducted in violation of its inspection agreement with the IAEA, Iran was forced to provide new information on this work and to explain its purpose. Iran's explanations, along with the results of the IAEA's inspections, were published in a series of reports beginning in June 2003. Yet, Iran has steadily maintained a course of domestic capacity building.

Towards the end of that year (2003), the foreign Ministries of France, Germany, and the UK visited Tehran to discuss their concerns about Iranian nuclear issues and voluntarily signed the additional non-proliferation protocol without being obligated under the terms of NPT to do so. With the signing of this agreement, Iran aimed to eliminate any misperceptions concerning its nuclear program. The European signatories of the Tehran accord in turn, agreed to first explicitly recognize Iran's rights to conduct nuclear research, and second, to discuss methods by which Iran could provide satisfactory assurances about the peaceful nature of its nuclear research program. Iran met these requirements, and the European countries were then to provide Iran with easy access to advanced nuclear technology. However, as a result of the European powers' delay in providing Iran with access to nuclear technology, and their adoption of a policy of marking time to impede Iranian research and development of nuclear technology, Iran, in Feb. 2005, requested that the European negotiators speed up the negotiation process, a request that was refused. Facing European intransigence, and shortly after the election of president Ahmadinejad, in August 2005, Iran had no alternative but to resume its nuclear activities by breaking the IAEA's seals on the equipment in the UCF nuclear facilities in Isfahan. Thus, a new era of Iranian R&D in nuclear technology began.

2. 2005-2013: In September 2005, the board of governors of IAEA, based on its 2003 report, declared that Iran had not complied with its safeguards agreement and voted to report Iran to the UN Security Council. This vote set the stage for a number of Security Council resolutions against Iran, sponsored by the UK, Germany and France, and with the backing of the US. Iran, in response, suspended all the voluntary cooperation beyond that which the country was obligated to observe with the IAEA, including implementation of the additional protocol. In April 8th 2006, President Ahmadinejad officially announced Iran's new technological capabilities for enriching uranium, as well as the establishment of a complete chain of uranium enrichment centrifuges in Natanz. 9 February 2010, Iran announced that it would enrich uranium up to 20 per cent to create fuel for a research reactor that used for producing medical radioisotopes, adding toits existing stocks of 3.5 per cent enriched uranium. Of course, making this announcement during afifth round negotiations with the E3+3 in fact did not help the already highly charged atmosphere and escalated the tensions and disputes.

3. In this period, the executive branch was dominated by the conservative advocacy coalition. Because, the policy belief of this coalition is vested in an inflexible position regarding the policy paradigm as well as policy instruments and settings, the acceptability-set of the statesmen is much smaller than the national win-set. During this period, the country's top negotiators were Ali Larijani and MohammdJalaili whose stances were both hardliners.

4. At long last, in November 2013, the diplomacy began to function, and against a backdrop of rising tensions and bellicose rhetoric, sanity prevailed and the world suddenly changed. Subsequently, a brand new chapter in Iran's relation with P5+1 (the five permanent members of the Security Council and Germany) emerged, characterized by some as a harbinger for hope and peace.

The current period started from the election of Mr. Rouhani in 2013. Since the position of Rouhani administration regarding the nuclear policy and diplomacy looks more like a policy broker than a policy advocate, mediating between two devil shift like competing coalitions and helping them realign through an evidence based policy making, it is expected that the domestic win-set and the acceptability-set of the statesmen correspond almost perfectly.

In the case of an administration seeking maximum flexibility, the acceptability–set lies at least partially outside the domestic win-set and closer to the opposing countries' win-set and in the case of an administration driven by zero-tolerance flexibility, the acceptability–set lies at least partially outside the domestic win-set but further from the opposing countries' win-set than the set of the agreement that can be ratified domestically. In the case of an administration operating as a policy broker intent on realigning the two competing coalitions, there is no conflict or even discrepancy between the negotiating teams and the society's acceptance of a deal.

We end this section with this conclusion which in each of these three periods, a specific configuration of the statesmen acceptability–set and the national win-set, and consequently a different strategy and bargaining power, has been articulated. The current domestic win-set is materialized by the introduction of an evidence-based policy making in the nuclear technology policy domain, making the alignment of the two coalitions within subsystem possible. As Evans describes, "The statesmen acceptability-set reflects the interests of the median domestic group and is encompassed by the domestic win-set if the statesmen were agents".

## 4. Conclusions

The nuclear technology policy of Iran has a hard foundation of the defense of the national rights created by the membership of the NPT. This underlying foundation is shared by all of the political factions and groups of political elites. However, when it comes to the working strategy on how Iran practically enjoys the right to attain the economic growth, and employing the specific policy tools, there are big disagreements among the political groups. We can distinguish roughly at least two advocacy coalitions: one which has zero tolerance for flexibility in policy instrument choice, and the other which has maximum flexibility for the choice of policy instruments. These two parties, locked in conflict for a long time, have now started to align their policy positions regarding the policy instruments. Consequently the aggregate of the acceptability sets of both advocacy coalitions has widened the national win-set.

This case study guided the paper to build a hybrid model integrating the domestic S&T policy and international diplomacy. According to this model this is competing belief systems which shape the paradigm shifts of policies and in turn they determine the change of diplomatic win-set in the international arena.

In line with this model, policy coalition who advocates a specific belief systems regarding an international hot issue, in the form of positive sum game, has to face this harsh reality that its acceptability–set lies at least partially outside the domestic win-set, and closer to the opposing countries' win-set and in the case of an administration driven by zero-sum game, the acceptability–set lies at least partially outside the domestic win-set but further from the opposing countries' win-set than the set of the agreement that can be ratified domestically.

Accordingly, it is only in the case of an administration operating as a policy broker intent on realigning the two competing coalitions, in which there is no conflict or even discrepancy between the negotiating teams and the society's acceptance of a deal.

It is clear that the win-set, the policy package acceptable and rectifiable by the internal political powers, does not have a fixed content. It has a dynamic and sometimes fluid content based on different factors. Evans has emphasized on the capacity of manipulation by statesmen. The author, however, add the role of trans-subsystem learning and accumulation of technical information absorbed by the members of both competing coalitions. Moreover, the role of enlightenment factor could not be overstressed. The former can increase the analytical capacity while the increase of the latter is attributed to the growth of reflective capacity. Both are very much needed for any process of evidence-based policy making.

This model is just a rudimentary step to explain the relation between two highly specialized research domains. Without any doubt, it has to be completed and revised by other theoretical and applied studies and the other researches on the other fields of science and technology policy.

## 5. Acknowledgements

This paper is the result of a research project founded by Iranian Science Foundation which I would express my gratitude.

## REFERENCES

1.  Baca, F. C. (n.d.). advocacy coalition framework; a theoretical framework for SAMREMto address policy change and learning cultivating community. capital institute de estudios ecuatorians quito equator Iwoa state.

2.  Borras, S. (2011). Policy learning and organizational capacity. Science and Public Policy, 38(9)November, 725–734.

3.  BrattDuane. (2012). Canada, the Province, and the Global Nuclear Revival: Advocacy Coalitions in Action. Montreal: McGill-Queens's Press.

4.  Chubin, S. (2006). The Iranian Nuclear Ambitions. Washungton D.C.: Carnegie Endowment for International Peace.

5.  Chubin, S. (2006). The Iranian Nuclear Ambitions. Washungton D.C.: Carnegie Endowment for International Peace.

6.  Cohen Roger (2014), Iranian Reality Check, New York Times

7.  Entessar, N. (2009, SummerVolume XVI, Number 2). Iran Nuclear Decision Making Calculus. Middle Eastern Policy Council.

8.  Evans Peter, Harold K. Jacobson, Robert Putnam (1993), Double-Edged Diplomacy, International Bargaining and Domestic Politics, Berkeley University,USA

9.  Gourevitch. (1978). The Second Image Reversed: the International Sources of Domestic Politics" (IO 32:4,Autumn 1978)

10. Hall, P. (1993). Policy Paradigm, Social Learning and the State. Comparative Politics, 273-296.

11. Katzenstein. (1983).

12. Lee, S. M. (2014). Understanding the Yalta Axioms and Riga, Axioms through beleif systems of the Advocacy Coalition Framework. Foreign Policy Analysis,0, 1-21.

13. Leonard, M. (2005). Can EU diplomacy stop the Iran's nuclear program? London: The Center for European Reform.

14. LiftinKaren. (2000). Advocacy Coalitions Along the Domestic-Foreign Frontier: Globalization and Canadian Climate Change Policy.

15. Michael Howlett, M. (2002). The Policy Effects of Internationalization:. Journal of Comparative Policy Analysis: Research and Practice 4: 31–50, 30-50.

16. NikouSamira. (N/A). Timeline of Iran's Nuclear Activities. http://iranprimer. usip.org/sites/iranprimer.usip.org/files/Timeline per cent 20of per cent 20Iran_s per cent 20Nuclear per cent 20Activities.pdf: United Institute for peace.

17. Olivier, M. (2006). Iran and Foreign Enrichment: A Troubled Model. Washington D.C.: Arms Control Today, https://www.armscontrol.org/act/2006_01-02/ JANFEB-IranEnrich, visited 4/15/2014.

18. Peter Evans, H. K. (1993). Double Edge Diplomacy, International Bargaining and Domestic Politics. Berkely: university of California Press.

19. Rosenaujames. (1969). Linkage politics: essays on the convergence of national and international systems. Free Press.

20. Rouhani, H. (2011). National Security and Nuclear Diplomacy, A memoir. Tehran: Center for Strategic Studies.

21. Sabatier, P. (1993). Policy Change and learning : An Advocacy Coalition Approach. Westview: University of Colorado.

22. Sabatier, P. (1998). Sabatier, P. A. (1998). The advocacy coalition framework : revisions and relevance for Europe. Journal of European public policy, 5(1), 98-130.

23. Sabatier, P. (1999). Theories of the policy process. Boulder, CO : Westview. Westview: Colorado University.

24. Sabatier, P. (2001), In N. J. Smelser and B. Baltes (eds.) *International Encyclopedia of the Social and Behavioral Sciences*. 17—11563 (2001).

25. Sich, A. (2012). How Iran Risks Another Chernobyl. http://thediplomat. com/2012/03/how-iran-risks-another-chernobyl/.

26. TabtabiiArian. (2013). Presidential Elections and Nuclear Policy In Iran. https:// www.armscontrol.org/act/2013_06/Presidential-Elections-and-Nuclear-Policy-In-Iran: arms control today.

## Chapter 14

# The Turkish Vision for Science, Technology, and Innovation

*Siir Kilkis[1] and Nesibe Yazıcı[2]*

*The Scientific and Technological Research Council of Turkey (TÜBITAK)*
*Department of Science, Technology, and Innovation Policy, Ankara, Turkey*
*e-mail: [1]siir.kilkis@tubitak.gov.tr, [2]nesibe.yazici@tubitak.gov.tr*

## ABSTRACT

The goal of sustainable development emphasizes the need to decouple economic growth from environmental degradation while addressing societal needs to increase well-being. Innovation systems are a key driver of sustainable development. In Turkey, the National Science, Technology, and Innovation Strategy (UBSTY) 2011-2016 define priority areas that are vital for the sustainable development of the country. These include energy, water, food, health, information and communication technology (ICT), machinery and manufacturing, automotive, defense, and space. In addition to being the main strategy document that guides the future of the innovation system, it is the first strategy that integrates national and international STI strategies. This paper begins by providing an overview of the increasingly more mature and vibrant R&D, innovation, and entrepreneurship system of Turkey as the basis of advancing opportunities to increase international cooperation. Such an overview is based on a unique application of the "functional dynamics" approach in the literature to characterize the Turkish innovation ecosystem. These functions are facilitating experimentation and learning, knowledge development, knowledge diffusion, guidance of search and selection, market formation, and the development and mobilization of resources. The paper proceeds by presenting developments in fostering international cooperation in areas of mutual interest, which includes the status of bilateral R&D cooperation, multilateral cooperation in programs and platforms, international scholarships and fellowships for incoming mobility of students and researchers, R&D and innovation study visits to emerging economies, Leadership Schools for senior managers of visiting countries, and international aid initiatives for Least Developed Countries. The paper concludes with perspectives on science diplomacy for sustainable development and a proposal for "innovation diplomacy" based on increased efforts for international function-to-function relations.

*Keywords:*     *Innovation systems, Functional dynamics, Science/Innovation diplomacy, Sustainable development.*

# 1. Introduction

The Turkish R&D, innovation, and entrepreneurship system is becoming an increasingly more mature and vibrant ecosystem. By the year 2023, which marks the 100th anniversary of the Republic of Turkey, the target is to increase R&D spending to 3 per cent of GDP. It is expected that the private sector R&D spending will be 2 per cent of GDP. In addition, the number of full-time equivalent (FTE) researchers is targeted to reach 300 thousand, which will support the increases in R&D spending from the dimension of human resources. Figure 14.1 puts forth the current progress of Turkey regarding these targets. Turkey further realizes that developing knowledge-intense technologies and creating high-value products have an integral role in escaping the "middle-income trap," which limits the increase of human welfare relative to increases in GDP. The critical solution pathway is to build an "innovation-based economy" through strengthening the national innovation ecosystem and increasing international cooperation through science diplomacy. An analysis of the countries that are innovation-based economies indicates a higher global competitiveness ranking score. In addition, sustainable development is a prerequisite for long-term competitiveness in this century.

This paper overviews the recent developments of Turkey in mobilizing science, technology, and innovation (STI) as the main drivers toward a more sustainable future and increasing vital international collaborations for R&D. The paper is structured into an overview of the national innovation system of Turkey based on the methodology of the "functional dynamics approach." The results and discussions overview the National Science, Technology, and Innovation Strategy (UBTYS) 2011-2016, which seeks to increase the functions of the innovation system. This strategy integrates national and international STI strategies for the first time. This section also overviews the dedicated efforts of Turkey for international R&D and innovation cooperation through science diplomacy. The paper concludes with a proposal for "innovation diplomacy" to increase function-to-function relations between the innovation systems of countries for sustainable development.

This paper overviews the recent developments of Turkey in mobilizing science, technology, and innovation (STI) as the main drivers toward a more sustainable future and increasing vital international collaborations for R&D. The paper is structured into an overview of the national innovation system of Turkey based on the methodology of the "functional dynamics approach." The results and discussions overview the National Science, Technology, and Innovation Strategy (UBTYS) 2011-2016, which seeks to increase the functions of the innovation system. This strategy integrates national and international STI strategies for the first time.

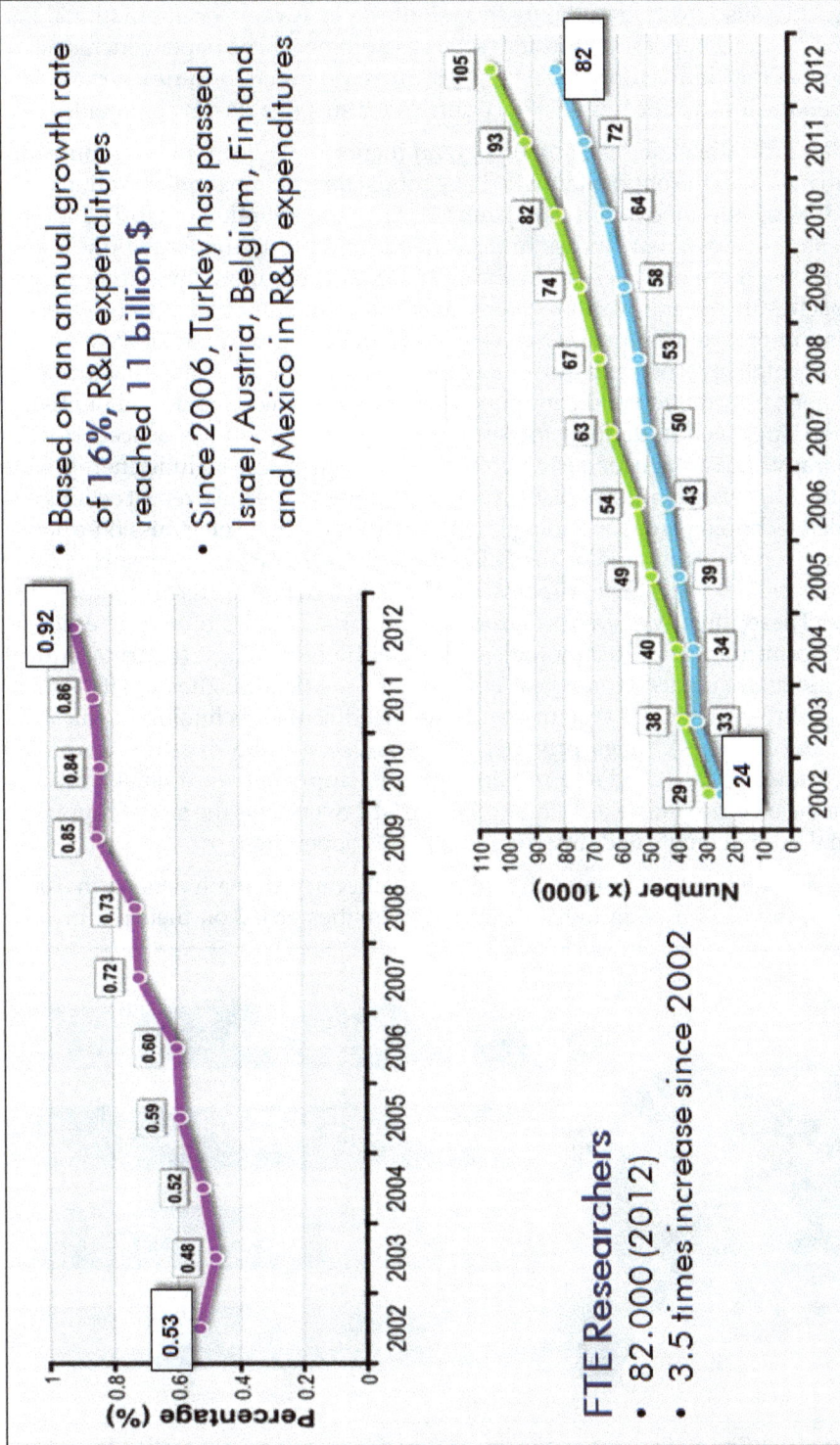

- Based on an annual growth rate of 16%, R&D expenditures reached 11 billion $

- Since 2006, Turkey has passed Israel, Austria, Belgium, Finland and Mexico in R&D expenditures

FTE Researchers
- 82.000 (2012)
- 3.5 times increase since 2002

**Figure 14.1: Progress in R&D Expenditures as a Share of GDP (Top) and FTE Researchers (Bottom).**

This section also overviews the dedicated efforts of Turkey for international R&D and innovation cooperation through science diplomacy. The paper concludes with a proposal for "innovation diplomacy" to increase function-to-function relations between the innovation systems of countries for sustainable development.

Based on the experience of Turkey, an increasingly more mature innovation system increases the opportunities to have robust international collaborations in the field of R&D and innovation. For example, since the adoption of UBTY 2011-2016 and a series of detailed studies to prioritize economic sectors based on multiple dimensions, there has been an evolving trend towards more "mission-oriented" approaches in the innovation system and international cooperation. One kind of mission-oriented approach has been to target sectors in which Turkey has a relatively high level of comparative advantage (*i.e.* automotive, manufacturing and machinery, information and community technology).The second kind of mission-oriented approach is defined for sectors in which Turkey has a need to attain comparative advantage for sustainable development. These include energy, water, food, health, defense, and space. In this respect, some of the most recent actions have included technology road mapping initiatives to support the new mission-oriented, call-based grant programs of the Scientific and Technological Research Council of Turkey (TÜB TAK). This process involved five stages to receive stakeholder inputs. These initiatives were conducted in the pilot topics of energy efficiency, mobile communication technologies, medical biotechnology (pharmaceuticals, vaccines, biomaterials, biomedical equipments, medical diagnostic kits), micro/ nano electro-mechanical systems, and advanced screen technologies (including OLED technologies). Mega projects, which are seen to be flagship "landmark" projects are also launched as a mission-oriented approach, most notably towards promising electric vehicle technology. Figure 14.2 overviews the aim of topping-off national achievements with international R&D cooperation.

In a related aspect, one of the other novel policy domain actions in Turkey has been the launch of an index to rank universities based on their performance

**Figure 14.2: The Role of International R&D and Innovation Cooperation in Turkey.**

in innovation and entrepreneurship. This index, namely the "Entrepreneurial and Innovative University Index," is a means to create positive competition between universities and recognize those universities that satisfy their "third role" for engagement in the entrepreneurship ecosystem in addition to their traditional roles for education and research. Of the 23 indicators in the 5 dimensions of the index, one measures the level of internationalization in universities under the dimension of "Cooperation and Interaction." Other aspects of improving the innovation ecosystem from the side of the universities and human resources have included better scholarship structures for both inbound and outbound mobility and new grant mechanisms to promote multi-sectoral collaboration. Regarding R&D infrastructure, a pilot framework that contains both quantitative and qualitative measures to assess the performance of national, thematic, and central R&D centers in Turkey has been designed and implemented to a select sample of research centers. The methodology included integration with field visits to evaluate their performance. International collaboration is expected to receive higher levels of performance. In addition, Turkey is in the process of evaluating additional policy options to increase the depth of research that research centers are undertaking towards addressing the needs of the industry. Means of attracting the R&D centers are also being designed to stimulate basic research in the ecosystem.

## 2. Materials and Methods

### The National R&D, Innovation, and Entrepreneurship System of Turkey

As an system that is composed of an interconnected network of funding, policy-making, and performing bodies, the national research and development (R&D), innovation, and entrepreneurship system of the Republic of Turkey is becoming an increasingly more mature and vibrant system. Figure 3 provides an analytical construction of the national innovation system of Turkey based on the distribution of the actors that are involved in contributing to various activities in the innovation system. Such a mapping of the institutions is a unique adaptation of the "functional dynamics approach" that is introduced in the literature to define the main activities for well-functioning innovation systems (Hekkert, 2008; Bergek, 2009; OECD, 2010). These main activities represent of set of six main functions (F1-F6) that includes facilitation of experimentation and learning (F1), knowledge development (F2), and knowledge diffusion (F3). The main activities continue to include guidance of search and selection (F4), market formation (F5), and the development and mobilization of resources (F6). As a whole, a national innovation system needs a systemic working of all six functions with maximum synergy and minimum failures within and between functions. Each of the six main activities F1-F6 will now be overviewed in the context of Figure 14.3.

### Facilitation of Experimentation and Learning (F1)

An innovation system needs a fresh inflow of ideas with the potential of leading to new products, processes, and services for the benefit of the country and

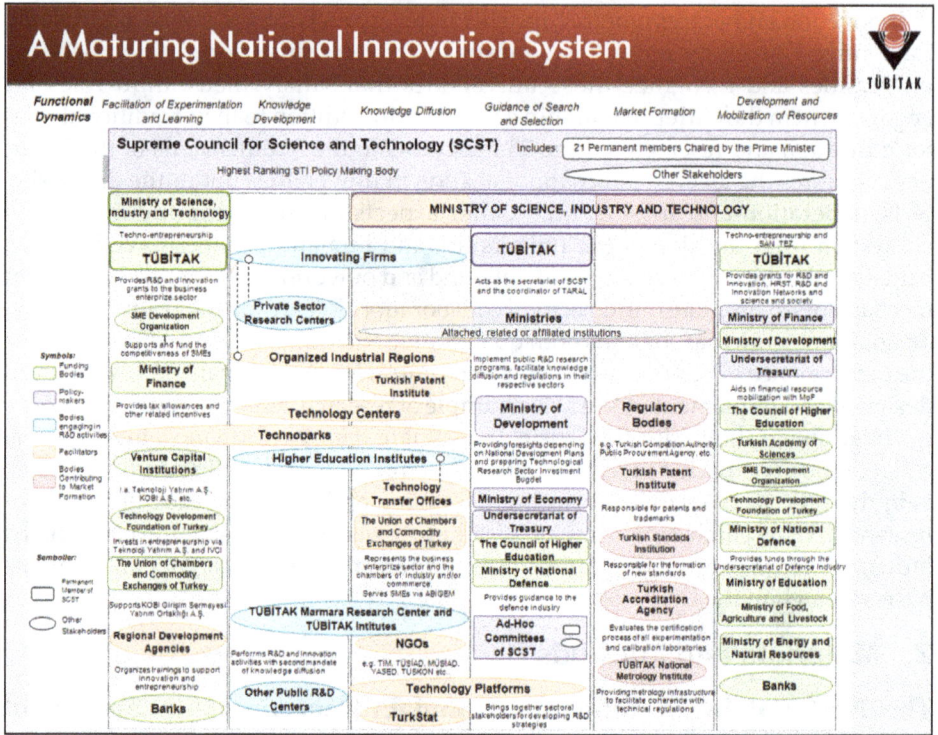

**Figure 14.3: Characterization of the National Innovation System of Turkey based on Functional Dynamics.**

humanity. For this reason, entrepreneurs must be motivated to pursue their ideas through the prototype stage and enter into the market with innovative products and technologies.

There are three public institutions that provide funding to support entrepreneurship activities in Turkey. These are the Ministry of Science, Industry, and Technology (BSTB/MoSIT), the Scientific and Technological Research Council of Turkey (TÜB TAK), and the SME Development Organization (KOSGEB). Each of these institutions implements direct support programs to support entrepreneurial activity (BSTB, TÜB TAK, and KOSGEB). An example may be given from the "Individual Entrepreneurship Staged Support Program" of TÜB TAK (Program 1512) that contains four phases to guide entrepreneurs from the stage of the idea to the stage of market entry and receiving venture capital as an R&D-intense start-up. There are also efforts to encourage start-ups to locate in international innovation hotspots, which include Silicon Valley and Bangalore.

BSTB is also responsible from following the performance of the Technology Development Zones and Techno-Parks in Turkey, which are important sources of innovative and entrepreneurial activity. The Ministry of Finance (MB) is responsible for implementing "Law 5746 on the Promotion of R&D Activities" that provides R&D based tax incentives and capital for techno-entrepreneur support programs.

Venture capital firms that support start-ups with innovative ideas, the Technology Development Foundation (TTGV), the Union of Chambers and Commodity Exchanges of Turkey (TOBB), Development Agencies under the Ministry of Development (KB/MoD), and banks are among the other actors with contributing roles to the functioning of F1.

## Knowledge Development (F2)

One of the most fundamental activities that are expected from an innovation system is the development of knowledge that is relevant to making both scientific and technological advancements in the present and potentially the future developments of the country. This requires a robust level of knowledge development in each of the R&D performing sectors, including higher education institutes, public research institutes, and the private sector.

In Turkey, higher education institutes, innovative firms, private sector research centers, technology parks, organized industrial zones, the TÜB TAK Marmara Research Center (MAM), and other public research institutes make important contributions to knowledge development. In addition, there exists a multitude of support mechanisms that are available for the benefit of these R&D and innovation actors either working singly or in collaboration as project partners in greater research consortiums. R&D performers are encouraged to undertake collaborative research activities while solving their own innovation related needs, their joint needs, or those of others. These may include a university that conducts R&D on behalf of an SME that has an R&D need or a consortium of universities and private sector firms that collaborate to satisfy the need of a public institution, such as a Ministry that exercises public procurement of R&D and innovation. There also exist programs to integrate a dimension of international collaboration into R&D projects, such as TÜB TAK 1011, and to carry out jointly-funded R&D projects under the bilateral agreements, including India.

## Knowledge Diffusion (F3)

The process of knowledge diffusion is a complex process that includes multi-directional feedbacks between multiple sources in the innovation system. Healthy flows and diffusion of knowledge sources, especially between the university and industry sectors, increase the chances that research results are transformed into societal, environmental, and economic benefit. Knowledge diffusion can be promoted through both collaborative R&D activities or through interfaces that facilitate the circulation of innovation-relevant knowledge between sectors.

A special support program of TÜB TAK (1513) supports the establishment and/or renewal of Technology Transfer Offices (TTO) at universities. Through this program, TTO's compete to receive grants to enhance the project support services that are offered to academicians, the portfolio management of university R&D projects and their transfer to the industry, and/or the collection of R&D needs from the local industry. TTO's can further request funds to include technology incubation facilities for faculty and university students who have an entrepreneurial spirit. Through the organization of project brokerage events and sectoral conferences, workshops, and fairs, sectoral Ministries further contribute to develop the climate of knowledge diffusion in Turkey.

The Turkish Patent Institute (TPE) is another actor in this respect based on the role of patents in diffusing knowledge about the status of state-of-the art technology in exchange for intellectual property protection. On the other hand, the Turkish Statistical Institute (TSE) collects, analyzes, and publishes data on indicators to track progress that is relevant for sustainable development, including R&D expenditures as a percentage of gross domestic product (GDP) and total FTE researchers.

### Guidance of Search and Selection (F4)

Innovation systems require the design of policies to improve both the "speed" of innovation (improving knowledge flows and collaborative interaction between the actors in the system) and the "direction" of innovation. The "direction" of innovation may include the selection and implementation of priority areas to provide focus for the R&D and innovation actors in the system. At the same time, it is important that the direction that is set for the priority areas allow room for creative ideas to continue to take place in the innovation system through technology-neutral, merit-based funding, including frontier research and curiosity-driven research of the R&D actors. From the experiences of Turkey, it is important that priority areas and their related topics are determined based on wide-ranging stakeholder participation from the sector and systemic policy instruments that are used to collect, analyze, deliberate, and rank priority areas.

The priority areas of UBTYS 2011-2016 are based on a series of stakeholder meetings and evaluations. These priority areas have been adopted by the Supreme Council for Science and Technology (SCST), which is the highest ranking decision-making body for the innovation system in Turkey. Following the adoption of the priority areas, high-level prioritization group meetings were used to determine the prioritized sub-fields of the priority areas. This effectively combined top-down direction setting through the SCST with the bottom-up direction setting of the innovation actors. In yet another level of stakeholder evaluations, TÜB TAK coordinates multi-staged processes based on the use of technology foresight methods, including Delphi surveys, to determine the prioritized topics under the prioritized sub-fields. Based on the timeline of the milestones in the technology roadmap, calls are opened under the mission-oriented programs of TÜB TAK, namely 1003 (for projects with university leadership) and 1511 (for projects with private sector leadership).

### Market Formation (F5)

The presence of "spaces" in the market to allow for the entry of innovative products, processes, and services depend on the balanced use of both supply-side technology-push policies, such as programs that fund pre-competitive R&D, and demand-side technology-pull policies, such as tax incentives, R&D and innovation based public procurement, and feed-in tariffs for renewable energy technologies. Some of the support mechanisms can have a hybrid design between supply-side and demand-side programs, such as uses of public procurement for R&D and innovation projects with industrial applications. TÜB TAK and some of the sectoral ministries, including the Ministry Transport, Maritime Affairs, and Communication (UDHB) have practices in this respect. Other actors that have a role in the process of market formation include the Public Procurement Agency (K K) and the Turkish Competition Authority (RK) of Turkey.

Due to the role of standards in defining technical qualifications for products and processes and the role of patents in protecting intellectual property in the market, TSE and TPE that have a role in knowledge diffusion also have a role in market formation. Other actors with roles under market formation include TÜB TAK National Metrology Institute (UME) that provides infrastructure to facilitate metrology tests with technical regulations and the Turkish Accreditation Agency (TÜRKAK), which evaluates the certification of experimentation and calibration laboratories.

### Development and Mobilization of Resources (F6)

The quality and quantity of human resources and R&D infrastructure are a key asset for R&D, innovation, and entrepreneurship activities. This requires an education system approach that focuses on the development of human resources from the primary education to the secondary and higher education levels. One of the themes of the SCST meetings have been on the "education system" in which decrees that relate to digital course content, structuring of scholarships, science centers, and science fairs were adopted (BTYK, 2012).

The main aim of these decrees is to allow the education system to reduce instances of memorization as much as possible and favor the use of analytical thinking skills to solve problems in a creative and knowledgeable way. The scholarships have also been restructured to provide additional incentives to students who study in a priority area, take education paths that have interdisciplinary degrees in the natural and social sciences, and/or declare a the pursuit of an academic degree in an area of industrial need. The scholarships for both outgoing and incoming students have also been developed to increase the mobility of students and international exchanges. Institutes that are responsible from student scholarships include TÜB TAK, KB, and the Council of Higher Education (YÖK). Under the San-Tez program of BSTB, students who undertake graduate theses to address an R&D need of the industry are given the opportunity to receive grants that are co-financed with the private sector. In addition, under the R&D tax incentive of Turkey, private sector firms that employ researchers with doctorate degrees receive deduction benefits from withholding income taxes. Regarding physical R&D infrastructure, KB provides grants to national, thematic, and central research laboratories at universities. There are also coordinated efforts with TÜB TAK to enhance the sustainability (human resources and funding) and R&D performance of research laboratories.

### Overview of the International Dimension of Functions

To be effective, all of the activities that are mentioned above have an international component. In fact, a robust and dynamic R&D, innovation, and entrepreneurship system allows Turkey to be more active in formulating and implementing international cooperation. The main aim of "science diplomacy" is to further these modes of cooperation between and among countries in meaningful ways for the countries that are involved.

## 3. Results and Discussion

### Overview of the International R&D Cooperation Dimension of UBTYS 2011-2016

The National Science, Technology, and Innovation Strategy (UBTYS) 2011-2016 is the main strategy document that is dedicated to setting forth a guiding direction for the future of the innovation system of Turkey. UBTYS 2011-2016 was adopted by decree no. 2010/201 of the Supreme Council for Science and Technology (SCST) as the highest ranking STI decision-making body in Turkey (BTYK, 2010). The strategy provides nine main domains of actions with three vertical dimensions, which define the priority areas for R&D in addition to curiosity-driven research, and six horizontal dimensions for capacity advancement. Figure 14.4 provides the main strategic framework of UBTYS with an overview of its main vertical and horizontal dimensions. This is the first strategy document in which national and

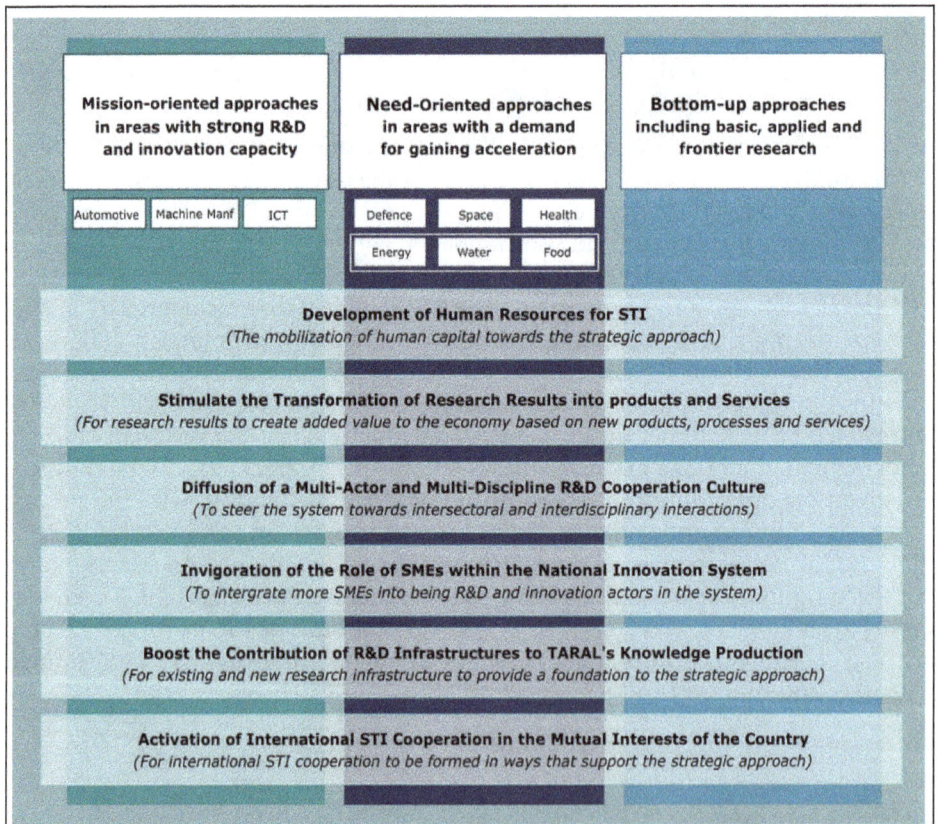

**Figure 14.4: The Strategic Framework of the National Science, Technology, and Innovation Strategy.**

international STI strategies are integrated. In addition, UBTYS 2011-2016 is the first strategy document that emphases "science diplomacy."

Regarding the vertical dimensions in Figure 14.4, the first set of priority areas is defined by those areas in which Turkey has a relatively strong comparative advantage. These are machinery and manufacturing, ICT, and the automotive sector. The second set focuses on those areas in which there are a need to provide impetus to R&D activities considering the sustainable development of the country. These are energy, water, food, health, defense, and space. The focus on curiosity-driven research of the researchers' own choosing is maintained to keep a balance between targeted and technology-neutral approaches as well as frontier research within the innovation system.

The dimensions for capacity advancement start with a focus on fostering human resources for STI, entrepreneurship, and increasing the penetration of STI culture in society. The dimensions continue with the pressing need of accelerating the commercialization of research results (second horizontal dimension) and increasing the level of multi-actor and multi-sector collaboration culture in the innovation system (third horizontal dimension). The fourth and fifth horizontal dimensions focus on increasing the number of R&D and innovation oriented SMEs and boosting the contribution of research infrastructure to support the Turkish innovation system. As a final dimension that tops-off all of the other dimensions is the dimension that focuses on international STI cooperation. UBTYS 2011-2016 thus aims to increase international STI cooperation in areas of mutual interest.

The integrated structure of the UBTYS 2011-2016 framework places an emphasis on international STI cooperation to support the thematic priorities of the country. This includes priority areas that aim to support sustainable development, such as energy, water, food, and health. In a related aspect, one of the aims is to promote active exchange and interaction in supra-national governance mechanisms as well as to initiate and diffuse the practice of "science diplomacy" activities. The Strategy also expands the scope of international aid packages to include STI-related aid. This is another aspect that supports the Millennium and the Sustainable Development Goals (MDG/SDGs). Within the scope of science diplomacy, the main aim is to initiate or diffuse the practice of sending science diplomat or "Science and Technology Counselors" to pilot countries and creating better coordination with the most similar Commerce Counselors under the Ministry of Economy. In 2012, this aim was supported by a Protocol that was signed between the Ministry of Foreign Affairs and the Ministry of Science, Industry and Technology to develop the practice of science diplomat.

## Science Diplomat Assignments to Pilot Countries and the Year(s) of Science

The first science diplomat is sent to Germany and the year 2014 has been declared as the Turkish-German Year of Science (TÜB TAK, 2014). In this pilot implementation, both countries expect to advance bilateral scientific and educational cooperation. The range of the cooperation is expected to range from joint R&D projects and to determine joint priority areas and promoting human capital exchange. This is based on a Memorandum of Understanding that is signed between the ministries of both countries and counterpart institutions, including with the German Research Foundation. In light of these activities, national and international workshops and seminars are being organized by public institutions and universities across Turkey and Germany. These workshops and seminars are being supported by TÜB TAK capped at a grant support of 50.000 TL per event (TÜB TAK, 2014). Based on the mutual areas of interest, the broad themes that have been identified by both countries are ICT, production technologies, transport, health research, energy, food and agriculture, climate change, environmental technologies, and demographic change (TÜB TAK, 2014). The application for events are currently being continued to be accepted by the 2223-D call of TÜB TAK.

The first pilot implementation of science diplomats is expected to provide important insight into the active practice of science diplomacy. The launch of similar science diplomacy missions in other countries is expected to increase collaboration between research institutions, facilitate the evaluation of potential innovation areas, increase cooperation in applied research, and support young researchers to take part in the scientific activities between Turkey and other pilot countries.

## International R&D Cooperation Programs (Programs of Cooperation)

Especially in today's globalized world where STI challenges transcend individual innovation systems, bilateral and multilateral R&D cooperation are an integral part of Turkey's STI policies. TÜB TAK has bilateral programs with 31 countries and multilateral participation in a total of 22 organizations, which also involve active forms of science and innovation diplomacy. A broad scope of international collaboration allows the Turkish STI ecosystem to top off achievements in national strategies with key contributions to societal and global challenges. At the same time, it provides an opportunity to foster multidisciplinary research, improve research collaboration, and engage in large scale and long term projects. It is important that scientific efforts to overcome challenges at the national level are coupled with adequate international cooperation to be more effective.

### Bilateral R&D Cooperation Programs

The bilateral programs of TÜB TAK consist of R&D Protocols with institutions as well as inter-governmental agreements. Table 14.1 provides the bilateral programs that have been signed between TÜB TAK and similar institutions in other countries. The scope of these bilateral programs ranges from co-funding of joint research projects and organization of scientific meetings to research exchanges and delegation study visits. One of the examples of the bilateral programs is the "TÜB TAK-CSIR Cooperation Program" (code 2506), which supports the joint R&D projects of researchers from Turkey and India. The duration of the R&D projects can be at most 2 years and both short-term and long-term scientist exchanges are supported by grants. A Program of Cooperation (POC) has been signed with the Department of Science and Technology (DST) towards implementing modes of international cooperation, including joint R&D calls.

## Table 14.1: Bilateral Programs of TÜBITAK Based on Country, Institute, and Program Code

| Partner Country | TÜBITAK Bilateral Programs | Program Code |
|---|---|---|
| Azerbaijan | Azerbaijan Academy of Sciences (ANAS) Bilateral Program | 2542 |
| Belarus | National Academy of Sciences (NASB) Bilateral Program | 2503 |
| Belgium | Flemish Research Organization Bilateral Program | 2539 |
| Bulgaria | National Academy of Sciences (BAS) Bilateral Program | 2502 |
| China | Ministry of Science and Technology (MoST) Bilateral Program | 2533 |
| Czech Republic | Academy of Sciences of Czech Republic Bilateral Program | 2537 |
| France | National Scientific Research Center (CNRS) Bilateral Program | 2505 |
| France | France Ministry of Foreign Affairs Bosporus Program | 2509 |
| Germany | Germany Research Council (DFG) Bilateral Program | 2507 |
| Germany | Ministry of Education and Research (BMBF) Bilateral Program | 2525 |
| Germany | Ministry of Education and Research (BMBF) IntenC Program | 2527 |
| Germany | Ministry of Education and Research (BMBF) 2+2 Program | 2534 |
| Greece | General Secretary of Research and Technology (GSRT) Bilateral Program | 2520 |
| Hungary | National Office of Research and Technology (NKTH) Bilateral Program | 2522 |
| India | Council for Scientific and Industrial Research (CSIR) Bilateral Program | 2506 |
| Italy | National Research Council (CNR) Bilateral Program | 2504 |
| Italy | Ministry of Foreign Affairs Bilateral Program | 2524 |
| Mongolia | Mongolian Academy of Sciences (MAS) Bilateral Program | 2526 |
| Montenegro | Ministry of Science Bilateral Program | 2541 |
| Morocco | National Scientific and Technological Research Center Bilateral Program | 2543 |
| Pakistan | Ministry of Science and Technology (MoST) Bilateral Program | 2529 |
| Romania | Romanian Scientific Research Authority (ANCS) Bilateral Program | 2528 |
| Russia | Russia Basic Research Foundation (RFBR) Bilateral Program | 2532 |
| Slovakia | Slovakian Academy of Sciences (SAS) Bilateral Program | 2513 |
| Slovakia | Slovakian Academy of Sciences (SAS) Thematic Bilateral Program | 2540 |
| Slovenia | Slovenian Research Council (ARRS) Bilateral Program | 2508 |
| South Korea | National Research Foundation (NRF) Bilateral Program | 2523 |
| Tunisia | Ministry of Higher Education and Scientific Research Bilateral Program | 2510 |
| Ukraine | National Academy of Sciences of Ukraine (NASU) Bilateral Program | 2512 |
| Ukraine | Science, Innovation and Information Committee Bilateral Program | 2514 |
| USA | National Science Foundation Bilateral Program | 2501 |

**Multilateral R&D Cooperation Programs and Platforms**

Turkey is a member of a wide-range of scientific cooperation networks, including the NAM S&T Centre, COST (European Cooperation in Science and Technology), ESA (European Space Agency) and EMBC (European Molecular Biology Conference). In addition, Turkey is a member of regional organizations, including BSEC (the Black Sea Economic Cooperation Organization), and ECO (Economic Cooperation Organization) in addition to international organizations, such as UNESCO, OECD, and NATO. TÜB TAK coordinates the participation of Turkish scientists and delegates in the STI related activities and events of these multilateral centers and organizations. In addition, TÜB TAK takes part in the Joint Programming Initiatives (JPI) at the European level most of which address the theme of sustainable development. Table 14.2 provides the related JPIs based on the priority areas.

Table 14.2: Cross-Relations of National Priority Areas and
Joint Programming Initiatives Participation

| Priority Type | Priority Area | JPI Participation |
|---|---|---|
| **"Need-based"** | Water | Water Challenges for a Changing World |
| (Sustainable development) | Food | Agriculture, Food Security and Climate Change |
| | Health | Healthy Diet for a Healthy Life |
| | Health | Antimicrobial Resistance |
| | Health | Neurodegenerative Disease Research |
| **Competitive Advantage** | ICT, Automotive | UrbanEurope |
| | ICT | More Years Better Lives |

At the policy-level, Turkey is further active in the Steering Committee of the Strategic Energy Technologies (SET-Plan) that has the purpose of accelerating cost-effective, low carbon energy technology innovation for an "energy transition." The SET-Plan is expected to contribute to targets of reducing primary energy spending, increasing energy efficiency, and increasing the share of renewable energy in electricity production each by to 20 per cent by the year 2020 (EU, 2013). In addition, TÜB TAK coordinates the participation of Turkish stakeholders, including public institutions and universities, in "Technology Platforms" at the European level. Table 14.3 compares the priority areas of Turkey with the related Technology Platforms. Their Turkish stakeholders range from the National Food Technology Platform to the Ministry of Transport, Maritime Affairs and Communication to the TÜB TAK Marmara Research Institute, universities, and firms. These stakeholders have more than 90 board/partner memberships in these Technology Platforms.

In addition, Turkey is an active participant as an associate member in "Horizon 2020" after similar active participation in the Framework Programs. One of the ERA-Net projects has been the "New INDIGO Era-Net" (Initiative for the Development and Integration of Indian and European Research), which aimed to increase the participation and dialogue of Indian researchers in the Seventh Framework Program. Under New INDIGO, joint project calls were launched in the areas of biotechnology and health in which 8 of a total of about 20 projects involved Turkish researchers. Turkish researchers are also involved in the energy calls that were recently announced. The INNO INDIGO (Innovation Driven Initiative for the Development and Integration of Indian and European Research) is an ongoing project upto the year 2016 whose calls will also include the private sector.

**Table 14.3: Cross-Relations of National Priority Areas and Joint Programming Initiatives Participation**

| Priority Type | Technology Platform Participation |
|---|---|
| **"Need-based"** | **EPTP**European Photovoltaic Technology Platform |
| (Sustainable development) | **Photonics21** European Technology Platform for Photonics |
| ☆ Energy | **RHC**European Technology Platform on Renewable Heating and Cooling |
| ☆ Wate | **ECTB** European Construction Technology Platform |
| ☆ Foo | **E2B A** Energy Efficient Buildings Association |
| ☆ Health | **EWEA** European Wind Energy Association |
| | **WSSTP** Water Supply and Sanitation Technology Platform |
| | **Waterborne** Water Technology Platform |
| | **GAH** European Technology Platform for Global Animal Health |
| | **Food for Life** Technology Platform |
| **Competitive Advantage** | **ManuFuture**European Technology Platform on Manufacturing |
| ☆ Manufacturing | **EUROP** European Technology Platform on Robotics |
| ☆ ICT | **ERTRAC**European Road Transport Research Advisory Council |
| ☆ Automotive/Transport | **ERRAC**The European Rail Research Advisory Council |
| | **NESSI** Networked European Software and Services Initiative |
| | **NEM** Networked and Electronic Media |
| | **Net!Works** Technology Platform for Communications Networks and Services |

Last but not least, Turkey is one of the most active countries in EUREKA, especially the clusters on ICT technologies that are known as "CELTIC" and "CELTIC Plus" clusters. In all of these ways, Turkey has robust R&D and innovation collaborations within multilateral platforms.

## International Scholarships and Fellowships for Incoming Mobility

Another area of international collaboration to which Turkey pays great attention is the provision of scholarships and fellowships to facilitate the flow of incoming mobility of international students and researchers. These include the Research Scholarship Program for International Researchers (2216), Graduate Scholarship Program for International Students (2215), and Fellowships for Visiting Scientists and Scientists on Sabbatical Leave. Through these programs, students and researchers may conduct studies and research in the fields of Natural Sciences, Engineering and Technological Sciences, Medical Sciences, Agricultural Sciences, Social Sciences and Humanities.

**Research Scholarship Program for International Researchers (2216)**

Under this program, TÜB TAK grants fellowships for international highly qualified PhD students and young post-doctoral researchers to pursue their research in Turkey. The program aims to promote Turkey's scientific and technological collaboration with countries of the prospective researchers. Preference is given to candidates who demonstrate the potential to contribute significantly to Turkey's goal of international cooperation in scientific and technological development. The maximum duration for the fellowship is 12 months with a monthly stipend of 2.250 Turkish Liras. In addition, travel costs up to 1000 US $and health insurance premiums up to 1200 US $ may be partially covered. In case of approval of the TÜB TAK B DEB Steering Committee, an extra research grant of up to 5.000 Turkish Liras may be provided for consumable expenses (materials, etc.) for the proposed research.

**Graduate Scholarship Program for International Students (2215)**

Through the scholarship program 2215, TÜB TAK grants scholarships for international students seeking to pursue a graduate degree in Turkey in the fields of Natural Sciences, Engineering and Technological Sciences, Medical Sciences, Agricultural Sciences, Social Sciences and Humanities. Clinical sciences are beyond the scope of scholarship program. The monthly stipend is 1.500 Turkish Liras for MSc/MA students while it is Turkish Liras for PhD students. An additional tuition fee of up to 2.000 Turkish Liras and a monthly allowance for health insurance coverage is also provided for the award holder only.

**Fellowships for Visiting Scientists and Scientists on Sabbatical Leave (2221)**

TÜB TAK grants fellowships for international scientists/researchers who would like to give workshops/conferences/lectures, or conduct R&D activities in Turkey. The program aims to promote Turkey's scientific and technological collaboration with the respective countries of the prospective fellows. Based on the possible fields of the scholarship, a list of more than a 100 research centers and institutes are provided on the TÜB TAK website, including the specific topics of their research, to facilitate the search of a potential fellow for a host research center and institution in Turkey.

There are three types of visit that may be granted within this program. The first type is for visiting scientists/researchers with a short-term visit to Turkey (up to 1 month). These may include conducting workshops/conferences/seminars, giving tutorials/lectures, participating in R&D activities, and organizing technical meetings for scientific and technological collaboration. Such activities under POC in bilateral R&D cooperation may also be funded through this program. The second type is for visiting scientists/researchers on a long term visit to Turkey (up to 12 months). The long-term fellowships fund is available for conducting R&D activities and teaching graduate/undergraduate courses. As for the third type of support, this is for scientists and researchers who are on sabbatical leave from their home countries and are planning to come to a Turkish university or institution. For durations of between 3 to 12 months, scholars or academic staff on sabbatical leave may also be funded to conduct R&D and teaching graduate/undergraduate courses.

The beneficiaries may be fellows of any country, should have a PhD degree (or equivalent) or have at least five years of research experience, and should be invited by a hosting institution in Turkey, which can be universities, research institutions, or industrial companies with a R&D unit. Fellows on sabbatical leave should be invited for duration of at least 3 months. Regarding the monthly stipend, visiting scientists receive a fellowship of up to 3.000 US$ and visiting scientists on sabbatical leave receive up to 3.500 US$. The amount of the monthly stipend will be determined based on the academic titles of fellows. Overall, the duration of the fellowship can vary from one week to 12 months. Travel costs (round trip) and health insurance premiums are also provided.

In the context of the above programs, a total of 95 students and guest researchers from India have been supported by TÜB TAK as of May 2014. This includes 5 PhD students under Program 2216, a total of 30 researchers under Program 2215, and 60 visiting scientists under Program 2221.

## R&D and Innovation Study Visits to Knowledge-Based and Emerging Economies

One of the vital components of science diplomacy is the ability to create fruitful dialogue and mutual exchange between countries at a level of willingness and openness to compare and benchmark mechanisms that are being used in the innovation system to support R&D and innovation activities. Such dialogue opens the possibility for better understanding of innovation systems, similar institutions, and counterpart support mechanisms. At the same time, it presents a possibility to build and develop a more solid basis of collaboration in the future. In this philosophy, TÜB TAK delegates have started to undertake exploratory R&D and innovation study visits, including to knowledge-based economies, such as Singapore, and emerging economies from among BRICS (Brasil, Russia, India, China, and South Africa) and MIST (Mexico, Indonesia, South Korea and Turkey) countries. Meetings are organized with the main institutions of the innovation system.

The R&D and Innovation Study Visit to India takes place June 2-6, 2014 and includes meetings with the Ministry of Science and Technology Department of Science and Technology (DST), the Technology Development Board (TDB), the Department of Scientific and Industrial Research (DSIR), the Council for Scientific and Industrial Research (CSIR), the Ministry of Communications and Information Technology Department of Electronics and Information Technology (DeitY), the Ministry of Finance, the Ministry of Human Resource Development, and Invest India. In fact, there are similarities between the STI Policy 2013 of India (DST, 2013) and the UBTYS 2011-2016 in Turkey (BTYK, 2010). Similar to the aspiration of India to establish new public-private partnership structures, seed high-risk innovations through new mechanisms, and foster strategic relationships for a high technology-led path[1], Turkey aspires to the escape from the "middle-income trap"

---

1 The Science, Research and Innovation System for High Technology-led path for India (SRISHTI) seeks to accelerate the pace of discovery and delivery of science-led solutions for faster, sustainable, inclusive growth (DST, 2013).

based on mission-oriented programs, knowledge-intense technologies, and high value creating products.

As a result of the R&D and innovation study visits, it is seen that countries, with their similarities and differences in economic structure and development goals, can effectively learn from each other in identifying best practices to promote research and technology development and foster a vibrant entrepreneurship ecosystem. Such an exchange also holds importance from the perspective of sustainable development since it is the common objective of countries to accelerate R&D and innovation to increase the capacity to decouple economic growth from environmental degradation.

## Organization of International Leadership Schools on Science and Technology Policy

As one of the institutes of TÜB TAK, the Turkish Institute for Industrial Management (TÜSS DE) has been organizing sessions on science and technology policy as "Leadership Schools" to experts and senior managers in developing countries. In these sessions, BSTB, TÜB TAK, KOSGEB, TPE, and TSE have been sharing their experiences and roles as actors of the Turkish innovation system (see Figure 14.2). These sessions include the sharing of experiences in science, technology, and innovation policies, R&D support mechanisms, SME support mechanisms, entrepreneurship policies, industrial strategies, organized industrial zones, patents, international and regional standards, and international cooperation. As an ongoing effort, the Leadership Schools have been organized for Central Asian, Balkan, and African countries, including Bosnia and Herzegovina, Kosovo, and Kazakhstan. Through these sessions, the participating experts and senior managers have obtained important insight into improving the R&D and innovation systems of their countries. In addition to the Leadership Schools, TÜB TAK officials have been meeting with senior level R&D managers of R&D institutes, including Morocco, Japan, Ecuador, Mexico, and Egypt.

## Scientific and Technological Opportunities for Least Developed Countries

In addition to providing leadership and training sessions, Turkey is involved in the Economic and Technical Cooperation Package for Least Developed Countries (LDC). Turkey firmly believes that it is a collective and shared responsibility to help LDCs, not just as a moral and ethical imperative, but also because global peace and security is directly linked with global sustainable development (MFA, 2013). The "Preparatory Event for the 4th UN Conference on the Least Developed Countries Science, Technology and Innovation: Setting priorities and implementing policies for LDCs" has been hosted by Turkey in 2011. Following the LDC Conference, a number of measures have been launched in Turkey to support capacity-building in LDCs, including in the fields of science and technology. These measures constitute the "Economic and Technical Cooperation Package of the Turkish Government for LDCs." Within this scope, TÜB TAK has initiated the 2235 Program.

## Graduate Scholarship Programme for Least Developed Countries Program (2235)

TÜB TAK is designated to host the 2235 Program to provide two students from each LDC country with master or PhD education in Turkey. The requirements for receiving the scholarship are that the student must be under 27 years of age for MSc. and 30 years of age PhD tracks at the closing date of application. In addition, the student must be accepted for MSc. (with thesis) or PhD programmes (except for straight-to-PhD programmes) from universities in Turkey which are listed in the official website of TÜB TAK (www.2235.tubitak.gov.tr). The maximum duration of the scholarship for MSc. students is 3 years and for PhD students, it is 5 years. The planned duration of the graduate scholarship programme for LDCs is 10 years. The scholarship includes a personal maintenance allowance of 1.500 TL for MSc students and 1.800 TL for PhD students, a travel allowance, semester fees, and support for Turkish language learning. In this respect, up to one year of Turkish language courses are paid by TÜB TAK. The students are expected to graduate as successful students in their field.

## Proposal on Perspectives on Science Diplomacy for Sustainable Development

Based on the experiences of the Republic of Turkey, increasing scientific, technological, and innovative exchanges between countries in meaningful relations as science diplomacy require an integrated effort. Figure 14.5 puts these experiences in a framework that provides a perspective for integrating the realms of national innovation systems given at least two countries through science diplomacy for sustainable development. Here, more than one national innovation system (system A, system B, system C, etc.) present in the global arena has robust interactions among their systems in which the realm of innovation diplomacy is to enhance function-to-function relations, diversify collaboration opportunities, and strengthen dialogue on best practices in meaningful ways, such as sustainable development and/or furthering the MDG or SDG through R&D and innovation.

For example, the launch of international start-ups and the opening of international incubators in the global entrepreneurial "hotspots" (Startup Genome, 2013) could be an extension of the function of fostering entrepreneurial activity (F1) at the level of innovation diplomacy. The launch of joint R&D programs and project calls is an extension of the function knowledge development (F2). The organization of joint workshops and scientists exchanges is a key element of knowledge diffusion (F3). At the same time, activities that are related to policy learning and sharing, such as delegation study visits and training sessions, as well as the coming together of experts to identify common priorities is another horizon of the function of guidance of search and selection (F4). As for the function of market formation (F5), issues of patents, standards, and technology transfer already involve an international dimension. The role of innovation diplomacy may be less explicit for this function than other functions. In contrast, one of the most directly used forms of innovation diplomacy is under resource development (F6) based on the opportunities that are made available by international scholarships and fellowships. Joint collaboration

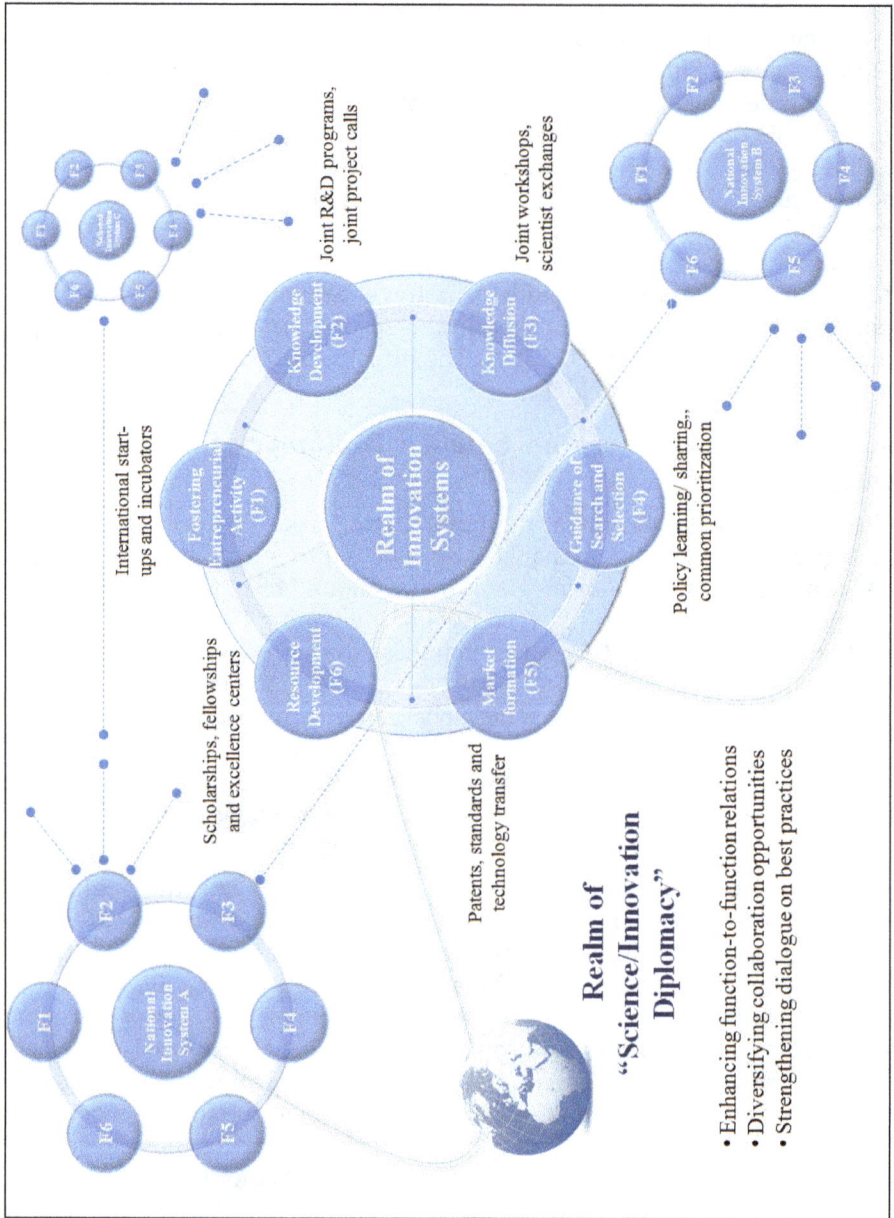

**Figure 14.5: Perspectives on Science Diplomacy for Sustainable Development Based on the Experiences and Vision of Turkey.**

activities in R&D infrastructure and excellence centers are among the possible extensions of F6 through innovation diplomacy.

From the scope of Figure 14.5, the possible domains of action for science diplomacy may be related to any one of the main activities F1-F6. As a result, the domains may be more diversified than purely scientific activities. For this reason, this paper proposes that it may be more fitting to refer to these activities as "innovation diplomacy" since it may encompass domains that relate to any extended activity of an innovation system, from promoting entrepreneurship to policy learning and joint R&D projects. The dotted lines in Figure 3 exemplify the possible modes of activities, some of which are overviewed below. These include bilateral R&D cooperation agreements with calls for joint funding of collaborative R&D projects, international scholarships and fellowships for incoming mobility, science diplomat assignment posts to pilot countries, "International Year(s) of Science," R&D and innovation study visits to knowledge-based and emerging economies, and the organization of international "Leadership Schools" on science and technology policy.

## 4. Conclusions

The Republic of Turkey has a robust foundation to pursue and further science diplomacy in the future. This is based on both an increasingly mature and dynamic R&D, innovation, and entrepreneurship ecosystem as well as the willingness of the country to further international cooperation in the areas of mutual interest for the benefit of humanity. This paper has overviewed the national innovation ecosystem of Turkey and the realm of "science and innovation diplomacy" to build greater ties for international cooperation in the field of science, technology, and innovation. Through all of these efforts, Turkey hopes to realize the guiding vision of UBTYS 2011-2016, which is *"To contribute to new knowledge and develop innovative technologies to improve the quality of life by transforming the former into products, processes, and services for the benefit of the country and humanity"* (BTYK, 2010). The sustainable development of our world depends on each country putting forth its best effort to progress towards a more clean, inclusive, and innovative future. International cooperation based on science and innovation diplomacy has an integral role in making this a reality that will be intricately linked with a more peaceful and sustainable future.

## 5. Acknowledgements

The authors acknowledge their institution, the Scientific and Technological Research Council of Turkey (TÜB TAK), for the information that was acquired for this paper during the career experience of the authors and the NAM S&T Centre for the opportunity of this Workshop.

## REFERENCES

1.  Bergek, A., Jacobsson S., Carlsson B., Lindmark S., Rickne A. (2008), Analyzing The Functional Dynamics Of Technological Innovation Systems: A Scheme Of Analysis, *Research Policy*, Vol. 37, pp. 407–429.

2. EU, (2013), SET Plan Secretariat, Scoping Paper Integrated Roadmap – As Proposed in the Communications on Energy Technologies and Innovation, Brussels, 13 June 2013.

3. Hekkert, M., Suurs, R., Negro, S., Kuhlmann, S., Smits, R. (2007), Functions of Innovation Systems: A New Approach for Analysing Technological Change, *Technological Forecasting and Social Change*, Vol. 74, No. 4, pp. 413–432.

4. Ministry of Foreign Affairs –MFA, (2013) <http://www.mfa.gov.tr/turkey_s-development-cooperation.en.mfa>.

5. OECD, (2010), Country Reports of Innovation Policy, Synthesis Report, Paris.

6. Startup Genome, (2013), Ranking of Entrepreneurship Hotspots.

7. Supreme Council for Science and Technology –BTYK/SCST, (2010), Decree No. 2010/101 for UBTYS 2011-2016, Ankara: TÜB TAK, 2010.

8. Supreme Council for Science and Technology –BTYK/SCST, (2010), Presentation of the President of Turkey at the 24th Meeting of the SCST Focused on the Education System: TÜB TAK.

9. TÜB TAK, (2014-a), Turkish-German Year of Science <http://www.tubitak.gov.tr/tr/duyuru/turk-alman-bilim-yili-etkinlik-destegi>.

10. TÜB TAK, (2014-b), Opening Ceremony of the 2014 Turkish-German Year of Science <http://www.tubitak.gov.tr/en/news/the-opening-ceremony-of-the-2014-turkish-german-year-of-science-was-held-in-berlin>.

# Manesar Declaration–2014
# Perspectives on Science and Technology Diplomacy for Sustainable Development in NAM and Other Developing Countries

**WHILE EXPRESSING GRATITUDE to** the Centre for Science and Technology of the Non-Aligned and Other Developing Countries (NAM S&T Centre) for hosting the International Workshop on 'Perspectives on Science and Technology Diplomacy for Sustainable Development in NAM and Other Developing Countries' at Manesar (Haryana), India during 27-30 May 2014 in partnership with the Department of Science and Technology (DST), Ministry of Science and Technology, Government of India;

**RECALLING the** 16th NAM Summit Declaration in Tehran, the Islamic Republic of Iran adopted in August 2012 recognising the importance of South-South cooperation in Science and Technology; and

**TAKING NOTE** of the Declaration adopted by the International Workshop on 'Science and Technology Diplomacy for Developing Countries', jointly organized during 13-15 May 2012 by the NAM S&T Centre and the Centre for Innovation and Technology Cooperation (CITC)-Presidency of the Islamic Republic of Iran in Tehran; and

**RECOGNISING that** Science and Technology is a global asset with the potential to contribute to inclusive socio-economic development, and to reinforce security in its various manifestations, such as energy, water, food and health;

**RECOGNISING also** the role being played by the UN agencies, inter-governmental bodies and various international organisations on Science, Technology and Innovation in raising awareness about Science and Technology Diplomacy;

**ACKNOWLEDGING** the important role that Science and Technology Diplomacy can play in achieving the Millennium Development Goals (MDGs) and the post-2015 development agenda;

**WE, THE PARTICIPANTS OF THE WORKSHOP**, representing Afghanistan, Cambodia, Colombia, Egypt, Germany, India, Indonesia, Iran, Malaysia, Mauritius, Myanmar, Nepal, Nigeria, Pakistan, Sri Lanka, South Africa, Switzerland, Syria, Turkey, Venezuela, Zambia and Zimbabwe; and

**HAVING DELIBERATED** on the perspectives of Science and Technology Diplomacy in the context of developing countries;

**UNANIMOUSLY** recommend:

☆ Augmenting the institutional and human capacity in Science and Technology Diplomacy for developing countries to achieve inclusive socio-economic development and their engagement in international discourse.

☆ Developing a framework for the adoption of Science and Technology Diplomacy as a tool for engagement of NAM and other developing countries to develop and strengthen their national S&T and innovation systems.

☆ Promoting Science and Technology Diplomacy as a distinct discipline by bringing out white papers, reports, policies and case studies, introducing postgraduate courses, research programmes and creating science diplomacy platforms for networking through North-South and South-South partnerships.

☆ Establishing appropriate fora to include science communication activities and heritage/indigenous knowledge relevant to Science and Technology Diplomacy.

☆ Strengthening the practice of Science and Technology Diplomacy amongst developing countries by posting Science and Technology Attaches/ Counsellors in the respective foreign missions.

☆ Establishing a Centre for Science and Technology Diplomacy in a developing country with due consideration of the initiatives made in Science and Technology Diplomacy by NAM and other developing countries.

**THUS, RESOLVED IN MANESAR (HARYANA), INDIA ON THIS DAY, 30th MAY 2014.**